Practice

Eureka Math®

Grade 4 Fluency

Published by Great Minds®.

Printed in the U.S.A.
This book may be purchased from the publisher at eureka-math.org.
10 9 8 7 6 5 4 3 2 1

ISBN 978-1-64054-607-3

G4-M1-M7-P/F-04.2018

Learn • Practice • Succeed

Eureka Math® student materials for *A Story of Units®* (K–5) are available in the *Learn, Practice, Succeed* trio. This series supports differentiation and remediation while keeping student materials organized and accessible. Educators will find that the *Learn, Practice,* and *Succeed* series also offers coherent—and therefore, more effective—resources for Response to Intervention (RTI), extra practice, and summer learning.

Learn

Eureka Math Learn serves as a student's in-class companion where they show their thinking, share what they know, and watch their knowledge build every day. *Learn* assembles the daily classwork—Application Problems, Exit Tickets, Problem Sets, templates—in an easily stored and navigated volume.

Practice

Each *Eureka Math* lesson begins with a series of energetic, joyous fluency activities, including those found in *Eureka Math Practice.* Students who are fluent in their math facts can master more material more deeply. With *Practice,* students build competence in newly acquired skills and reinforce previous learning in preparation for the next lesson.

Together, *Learn* and *Practice* provide all the print materials students will use for their core math instruction.

Succeed

Eureka Math Succeed enables students to work individually toward mastery. These additional problem sets align lesson by lesson with classroom instruction, making them ideal for use as homework or extra practice. Each problem set is accompanied by a Homework Helper, a set of worked examples that illustrate how to solve similar problems.

Teachers and tutors can use *Succeed* books from prior grade levels as curriculum-consistent tools for filling gaps in foundational knowledge. Students will thrive and progress more quickly as familiar models facilitate connections to their current grade-level content.

Students, families, and educators:

Thank you for being part of the *Eureka Math*® community, where we celebrate the joy, wonder, and thrill of mathematics. One of the most obvious ways we display our excitement is through the fluency activities provided in *Eureka Math Practice*.

What is fluency in mathematics?

You may think of *fluency* as associated with the language arts, where it refers to speaking and writing with ease. In prekindergarten through grade 5, the *Eureka Math* curriculum contains multiple daily opportunities to build fluency *in mathematics*. Each is designed with the same notion—growing every student's ability to use mathematics *with ease*. Fluency experiences are generally fast-paced and energetic, celebrating improvement and focusing on recognizing patterns and connections within the material. They are not intended to be graded.

Eureka Math fluency activities provide differentiated practice through a variety of formats—some are conducted orally, some use manipulatives, others use a personal whiteboard, and still others use a handout and paper-and-pencil format. *Eureka Math Practice* provides each student with the printed fluency exercises for his or her grade level.

What is a Sprint?

Many printed fluency activities utilize the format we call a Sprint. These exercises build speed and accuracy with already acquired skills. Used when students are nearing optimum profi ciency, Sprints leverage tempo to build a low-stakes adrenaline boost that increases memory and recall. Their intentional design makes Sprints inherently differentiated; the problems build from simple to complex, with the first quadrant of problems being the simplest and each subsequent quadrant adding complexity. Further, intentional patterns within the sequence of problems engage students' higher order thinking skills.

The suggested format for delivering a Sprint calls for students to do two consecutive Sprints (labeled A and B) on the same skill, each timed at one minute. Students pause between Sprints to articulate the patterns they noticed as they worked the fi rst Sprint. Noticing the patterns often provides a natural boost to their performance on the second Sprint.

Sprints can be conducted with an untimed protocol as well. The untimed protocol is highly recommended when students are still building confi dence with the level of complexity of the fi rst quadrant of problems. Once all students are prepared for success on the Sprint, the work of improving speed and accuracy with the energy of a timed protocol is often welcome and invigorating.

Where can I find other fluency activities?

The *Eureka Math Teacher Edition* guides educators in the delivery of all fluency activities for each lesson, including those that do not require print materials. Additionally, the *Eureka Digital Suite* provides access to the fl uency activities for all grade levels, searchable by standard or lesson.

Best wishes for a year filled with aha moments!

Jill Diniz

Jill Diniz
Director of Mathematics
Great Minds

Contents

Module 1

Module 2

Module 3

Module 5

Module 4 does not contain any Sprints or printed fluency components. All fluency activities in Module 4 can be found in the Teacher Edition and can be completed without paper.

Module 6

Module 7

Grade 4
Module 1

A

Number Correct: _______

Multiply and Divide by 10

1.	2 × 10 =		23.	__ × 10 = 100	
2.	3 × 10 =		24.	__ × 10 = 20	
3.	4 × 10 =		25.	__ × 10 = 30	
4.	5 × 10 =		26.	100 ÷ 10 =	
5.	1 × 10 =		27.	50 ÷ 10 =	
6.	20 ÷ 10 =		28.	10 ÷ 10 =	
7.	30 ÷ 10 =		29.	20 ÷ 10 =	
8.	50 ÷ 10 =		30.	30 ÷ 10 =	
9.	10 ÷ 10 =		31.	__ × 10 = 60	
10.	40 ÷ 10 =		32.	__ × 10 = 70	
11.	6 × 10 =		33.	__ × 10 = 90	
12.	7 × 10 =		34.	__ × 10 = 80	
13.	8 × 10 =		35.	70 ÷ 10 =	
14.	9 × 10 =		36.	90 ÷ 10 =	
15.	10 × 10 =		37.	60 ÷ 10 =	
16.	80 ÷ 10 =		38.	80 ÷ 10 =	
17.	70 ÷ 10 =		39.	11 × 10 =	
18.	90 ÷ 10 =		40.	110 ÷ 10 =	
19.	60 ÷ 10 =		41.	30 ÷ 10 =	
20.	100 ÷ 10 =		42.	120 ÷ 10 =	
21.	__ × 10 = 50		43.	14 × 10 =	
22.	__ × 10 = 10		44.	140 ÷ 10 =	

B

Number Correct: ________

Multiply and Divide by 10

Improvement: ________

1.	1 × 10 =	10		23.	__ × 10 = 20	2
2.	2 × 10 =	20		24.	__ × 10 = 100	10
3.	3 × 10 =	30		25.	__ × 10 = 30	3
4.	4 × 10 =	40		26.	20 ÷ 10 =	2
5.	5 × 10 =	50		27.	10 ÷ 10 =	1
6.	30 ÷ 10 =	3		28.	100 ÷ 10 =	10
7.	20 ÷ 10 =	2		29.	50 ÷ 10 =	5
8.	40 ÷ 10 =	4		30.	30 ÷ 10 =	3
9.	10 ÷ 10 =	1		31.	__ × 10 = 30	3
10.	50 ÷ 10 =	5		32.	__ × 10 = 40	4
11.	10 × 10 =	100		33.	__ × 10 = 90	9
12.	6 × 10 =	60		34.	__ × 10 = 70	1 *
13.	7 × 10 =	10		35.	80 ÷ 10 =	8
14.	8 × 10 =	80		36.	90 ÷ 10 =	9
15.	9 × 10 =	90		37.	60 ÷ 10 =	6
16.	70 ÷ 10 =	1		38.	70 ÷ 10 =	1
17.	60 ÷ 10 =	6		39.	11 × 10 =	110
18.	80 ÷ 10 =	8		40.	110 ÷ 10 =	10
19.	100 ÷ 10 =	10		41.	12 × 10 =	120
20.	90 ÷ 10 =	9		42.	120 ÷ 10 =	12
21.	__ × 10 = 10	1		43.	13 × 10 =	130
22.	__ × 10 = 50	5		44.	130 ÷ 10 =	30

A

Number Correct: _______

Multiply by 3

1.	1 × 3 =		23.	10 × 3 =	
2.	3 × 1 =		24.	9 × 3 =	
3.	2 × 3 =		25.	4 × 3 =	
4.	3 × 2 =		26.	8 × 3 =	
5.	3 × 3 =		27.	5 × 3 =	
6.	4 × 3 =		28.	7 × 3 =	
7.	3 × 4 =		29.	6 × 3 =	
8.	5 × 3 =		30.	3 × 10 =	
9.	3 × 5 =		31.	3 × 5 =	
10.	6 × 3 =		32.	3 × 6 =	
11.	3 × 6 =		33.	3 × 1 =	
12.	7 × 3 =		34.	3 × 9 =	
13.	3 × 7 =		35.	3 × 4 =	
14.	8 × 3 =		36.	3 × 3 =	
15.	3 × 8 =		37.	3 × 2 =	
16.	9 × 3 =		38.	3 × 7 =	
17.	3 × 9 =		39.	3 × 8 =	
18.	10 × 3 =		40.	11 × 3 =	
19.	3 × 10 =		41.	3 × 11 =	
20.	3 × 3 =		42.	12 × 3 =	
21.	1 × 3 =		43.	3 × 13 =	
22.	2 × 3 =		44.	13 × 3 =	

B

Number Correct: ______

Multiply by 3

Improvement: ______

1.	3 × 1 =		23.	9 × 3 =	
2.	1 × 3 =		24.	3 × 3 =	
3.	3 × 2 =		25.	8 × 3 =	
4.	2 × 3 =		26.	4 × 3 =	
5.	3 × 3 =		27.	7 × 3 =	
6.	3 × 4 =		28.	5 × 3 =	
7.	4 × 3 =		29.	6 × 3 =	
8.	3 × 5 =		30.	3 × 5 =	
9.	5 × 3 =		31.	3 × 10 =	
10.	3 × 6 =		32.	3 × 1 =	
11.	6 × 3 =		33.	3 × 6 =	
12.	3 × 7 =		34.	3 × 4 =	
13.	7 × 3 =		35.	3 × 9 =	
14.	3 × 8 =		36.	3 × 2 =	
15.	8 × 3 =		37.	3 × 7 =	
16.	3 × 9 =		38.	3 × 3 =	
17.	9 × 3 =		39.	3 × 8 =	
18.	3 × 10 =		40.	11 × 3 =	
19.	10 × 3 =		41.	3 × 11 =	
20.	1 × 3 =		42.	13 × 3 =	
21.	10 × 3 =		43.	3 × 13 =	
22.	2 × 3 =		44.	12 × 3 =	

A

Number Correct: _______

Multiply by 4

1.	$1 \times 4 =$	
2.	$4 \times 1 =$	
3.	$2 \times 4 =$	
4.	$4 \times 2 =$	
5.	$3 \times 4 =$	
6.	$4 \times 3 =$	
7.	$4 \times 4 =$	
8.	$5 \times 4 =$	
9.	$4 \times 5 =$	
10.	$6 \times 4 =$	
11.	$4 \times 6 =$	
12.	$7 \times 4 =$	
13.	$4 \times 7 =$	
14.	$8 \times 4 =$	
15.	$4 \times 8 =$	
16.	$9 \times 4 =$	
17.	$4 \times 9 =$	
18.	$10 \times 4 =$	
19.	$4 \times 10 =$	
20.	$4 \times 3 =$	
21.	$1 \times 4 =$	
22.	$2 \times 4 =$	

23.	$10 \times 4 =$	
24.	$9 \times 4 =$	
25.	$4 \times 4 =$	
26.	$8 \times 4 =$	
27.	$4 \times 3 =$	
28.	$7 \times 4 =$	
29.	$6 \times 4 =$	
30.	$4 \times 10 =$	
31.	$4 \times 5 =$	
32.	$4 \times 6 =$	
33.	$4 \times 1 =$	
34.	$4 \times 9 =$	
35.	$4 \times 4 =$	
36.	$4 \times 3 =$	
37.	$4 \times 2 =$	
38.	$4 \times 7 =$	
39.	$4 \times 8 =$	
40.	$11 \times 4 =$	
41.	$4 \times 11 =$	
42.	$12 \times 4 =$	
43.	$4 \times 12 =$	
44.	$13 \times 4 =$	

B

Number Correct: ______

Multiply by 4

Improvement: ______

1.	4 × 1 =	
2.	1 × 4 =	
3.	4 × 2 =	
4.	2 × 4 =	
5.	4 × 3 =	
6.	3 × 4 =	
7.	4 × 4 =	
8.	4 × 5 =	
9.	5 × 4 =	
10.	4 × 6 =	
11.	6 × 4 =	
12.	4 × 7 =	
13.	7 × 4 =	
14.	4 × 8 =	
15.	8 × 4 =	
16.	4 × 9 =	
17.	9 × 4 =	
18.	4 × 10 =	
19.	10 × 4 =	
20.	1 × 4 =	
21.	10 × 4 =	
22.	2 × 4 =	

23.	9 × 4 =	
24.	3 × 4 =	
25.	8 × 4 =	
26.	4 × 4 =	
27.	7 × 4 =	
28.	5 × 4 =	
29.	6 × 4 =	
30.	4 × 5 =	
31.	4 × 10 =	
32.	4 × 1 =	
33.	4 × 6 =	
34.	4 × 4 =	
35.	4 × 9 =	
36.	4 × 2 =	
37.	4 × 7 =	
38.	4 × 3 =	
39.	4 × 8 =	
40.	11 × 4 =	
41.	4 × 11 =	
42.	12 × 4 =	
43.	4 × 12 =	
44.	13 × 4 =	

A

Number Correct: ______

Find the Midpoint

#			
1.	0	10	
2.	0	100	
3.	0	1000	
4.	10	20	
5.	100	200	
6.	1000	2000	
7.	30	40	
8.	300	400	
9.	400	500	
10.	20	30	
11.	30	40	
12.	40	50	
13.	50	60	
14.	500	600	
15.	5000	6000	
16.	200	300	
17.	300	400	
18.	700	800	
19.	5700	5800	
20.	70	80	
21.	670	680	
22.	6700	6800	
23.	6000	7000	
24.	600	700	
25.	60	70	
26.	260	270	
27.	9260	9270	
28.	80	90	
29.	90	100	
30.	990	1000	
31.	9990	10,000	
32.	440	450	
33.	8300	8400	
34.	680	690	
35.	9400	9500	
36.	3900	4000	
37.	2450	2460	
38.	7080	7090	
39.	3200	3210	
40.	8630	8640	
41.	8190	8200	
42.	2510	2520	
43.	4890	4900	
44.	6660	6670	

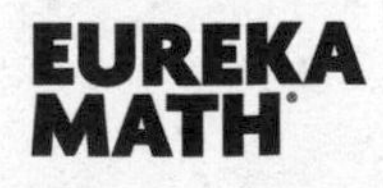

B

Number Correct: _______

Find the Midpoint

Improvement: _______

1.	10	20	
2.	100	200	
3.	1000	2000	
4.	20	30	
5.	200	300	
6.	2000	3000	
7.	40	50	
8.	400	500	
9.	500	600	
10.	30	40	
11.	40	50	
12.	50	60	
13.	60	70	
14.	600	700	
15.	6000	7000	
16.	300	400	
17.	400	500	
18.	800	900	
19.	5800	5900	
20.	80	90	
21.	680	690	
22.	6800	6900	

23.	7000	8000	
24.	700	800	
25.	70	80	
26.	270	280	
27.	9270	9280	
28.	80	90	
29.	90	100	
30.	990	1000	
31.	9990	10,000	
32.	450	460	
33.	8400	8500	
34.	580	590	
35.	9500	9600	
36.	2900	3000	
37.	3450	3460	
38.	6080	6090	
39.	4200	4210	
40.	7630	7640	
41.	7190	7200	
42.	3510	3520	
43.	5890	5900	
44.	7770	7780	

A

Number Correct: ______

Round to the Nearest 10,000

1.	21,000 ≈	20,000	23.	185,000 ≈	
2.	31,000 ≈	30,000	24.	85,000 ≈	
3.	41,000 ≈	40,000	25.	95,000 ≈	
4.	541,000 ≈	540,500	26.	97,000 ≈	
5.	49,000 ≈	50,000	27.	98,000 ≈	
6.	59,000 ≈		28.	198,000 ≈	
7.	69,000 ≈		29.	798,000 ≈	
8.	369,000 ≈		30.	31,200 ≈	
9.	62,000 ≈		31.	49,300 ≈	
10.	712,000 ≈		32.	649,300 ≈	
11.	28,000 ≈		33.	64,520 ≈	
12.	37,000 ≈		34.	164,520 ≈	
13.	137,000 ≈		35.	17,742 ≈	
14.	44,000 ≈		36.	917,742 ≈	
15.	56,000 ≈		37.	38,396 ≈	
16.	456,000 ≈		38.	64,501 ≈	
17.	15,000 ≈		39.	703,280 ≈	
18.	25,000 ≈		40.	239,500 ≈	
19.	35,000 ≈		41.	708,170 ≈	
20.	235,000 ≈		42.	188,631 ≈	
21.	75,000 ≈		43.	777,499 ≈	
22.	175,000 ≈		44.	444,919 ≈	

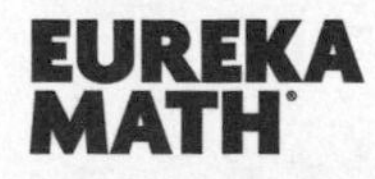

B

Number Correct: _______

Round to the Nearest 10,000

Improvement: _______

1.	11,000 ≈		23.	185,000 ≈	
2.	21,000 ≈		24.	85,000 ≈	
3.	31,000 ≈		25.	95,000 ≈	
4.	531,000 ≈		26.	96,000 ≈	
5.	39,000 ≈		27.	99,000 ≈	
6.	49,000 ≈		28.	199,000 ≈	
7.	59,000 ≈		29.	799,000 ≈	
8.	359,000 ≈		30.	21,200 ≈	
9.	52,000 ≈		31.	39,300 ≈	
10.	612,000 ≈		32.	639,300 ≈	
11.	18,000 ≈		33.	54,520 ≈	
12.	27,000 ≈		34.	154,520 ≈	
13.	127,000 ≈		35.	27,742 ≈	
14.	34,000 ≈		36.	927,742 ≈	
15.	46,000 ≈		37.	28,396 ≈	
16.	346,000 ≈		38.	54,501 ≈	
17.	25,000 ≈		39.	603,280 ≈	
18.	35,000 ≈		40.	139,500 ≈	
19.	45,000 ≈		41.	608,170 ≈	
20.	245,000 ≈		42.	177,631 ≈	
21.	65,000 ≈		43.	888,499 ≈	
22.	165,000 ≈		44.	444,909 ≈	

A

Number Correct: _______

Convert Meters and Centimeters to Centimeters

1.	2 m =	cm	23.	1 m 2 cm =	cm
2.	3 m =	cm	24.	1 m 3 cm =	cm
3.	4 m =	cm	25.	1 m 4 cm =	cm
4.	9 m =	cm	26.	1 m 7 cm =	cm
5.	1 m =	cm	27.	2 m 7 cm =	cm
6.	7 m =	cm	28.	3 m 7 cm =	cm
7.	5 m =	cm	29.	8 m 7 cm =	cm
8.	8 m =	cm	30.	8 m 4 cm =	cm
9.	6 m =	cm	31.	4 m 9 cm =	cm
10.	1 m 20 cm =	cm	32.	6 m 8 cm =	cm
11.	1 m 30 cm =	cm	33.	9 m 3 cm =	cm
12.	1 m 40 cm =	cm	34.	2 m 60 cm =	cm
13.	1 m 90 cm =	cm	35.	3 m 75 cm =	cm
14.	1 m 95 cm =	cm	36.	6 m 33 cm =	cm
15.	1 m 85 cm =	cm	37.	8 m 9 cm =	cm
16.	1 m 84 cm =	cm	38.	4 m 70 cm =	cm
17.	1 m 73 cm =	cm	39.	7 m 35 cm =	cm
18.	1 m 62 cm =	cm	40.	4 m 17 cm =	cm
19.	2 m 62 cm =	cm	41.	6 m 4 cm =	cm
20.	7 m 62 cm =	cm	42.	10 m 4 cm =	cm
21.	5 m 27 cm =	cm	43.	10 m 40 cm =	cm
22.	3 m 87 cm =	cm	44.	11 m 84 cm =	cm

B

Number Correct: _______

Convert Meters and Centimeters to Centimeters

Improvement: _______

1.	1 m =	cm	23.	1 m 1 cm =	cm
2.	2 m =	cm	24.	1 m 2 cm =	cm
3.	3 m =	cm	25.	1 m 3 cm =	cm
4.	7 m =	cm	26.	1 m 9 cm =	cm
5.	5 m =	cm	27.	2 m 9 cm =	cm
6.	9 m =	cm	28.	3 m 9 cm =	cm
7.	4 m =	cm	29.	7 m 9 cm =	cm
8.	8 m =	cm	30.	7 m 4 cm =	cm
9.	6 m =	cm	31.	4 m 8 cm =	cm
10.	1 m 10 cm =	cm	32.	6 m 3 cm =	cm
11.	1 m 20 cm =	cm	33.	9 m 5 cm =	cm
12.	1 m 30 cm =	cm	34.	2 m 50 cm =	cm
13.	1 m 70 cm =	cm	35.	3 m 85 cm =	cm
14.	1 m 75 cm =	cm	36.	6 m 31 cm =	cm
15.	1 m 65 cm =	cm	37.	6 m 7 cm =	cm
16.	1 m 64 cm =	cm	38.	4 m 60 cm =	cm
17.	1 m 53 cm =	cm	39.	7 m 25 cm =	cm
18.	1 m 42 cm =	cm	40.	4 m 13 cm =	cm
19.	2 m 42 cm =	cm	41.	6 m 2 cm =	cm
20.	8 m 42 cm =	cm	42.	10 m 3 cm =	cm
21.	5 m 29 cm =	cm	43.	10 m 30 cm =	cm
22.	3 m 89 cm =	cm	44.	11 m 48 cm =	cm

Grade 4
Module 2

Correct ______

Write in meters and centimeters.

1	3 m + 1 m =	m	cm	23	3 m 10 cm + 1 m 1 cm =	m	cm
2	4 m + 2 m =	m	cm	24	3 m 10 cm + 2 m 2 cm =	m	cm
3	2 m + 3 m =	m	cm	25	3 m 10 cm + 3 m 3 cm =	m	cm
4	5 m + 4 m =	m	cm	26	3 m 20 cm + 3 m 3 cm =	m	cm
5	2 m + 2 m =	m	cm	27	6 m 30 cm + 2 m 20 cm =	m	cm
6	3 m + 3 m =	m	cm	28	8 m 30 cm + 2 m 20 cm =	m	cm
7	4 m + 4 m =	m	cm	29	6 m 50 cm + 2 m 25 cm =	m	cm
8	5 m + 5 m =	m	cm	30	6 m 25 cm + 2 m 25 cm =	m	cm
9	5 m 7 cm + 1 m =	m	cm	31	4 m 70 cm + 1 m 10 cm =	m	cm
10	6 m 7 cm + 1 m =	m	cm	32	4 m 80 cm + 1 m 10 cm =	m	cm
11	7 m 7 cm + 1 m =	m	cm	33	4 m 90 cm + 1 m 10 cm =	m	cm
12	9 m 7 cm + 1 m =	m	cm	34	4 m 90 cm + 1 m 20 cm =	m	cm
13	9 m 7 cm + 1 cm =	m	cm	35	4 m 90 cm + 1 m 60 cm =	m	cm
14	5 m 7 cm + 1 cm =	m	cm	36	5 m 75 cm + 2 m 25 cm =	m	cm
15	3 m 7 cm + 1 cm =	m	cm	37	5 m 75 cm + 2 m 50 cm =	m	cm
16	3 m 7 cm + 3 cm =	m	cm	38	4 m 90 cm + 3 m 50 cm =	m	cm
17	6 m 70 cm + 10 cm =	m	cm	39	5 m 95 cm + 3 m 25 cm =	m	cm
18	6 m 80 cm + 10 cm =	m	cm	40	4 m 85 cm + 3 m 25 cm =	m	cm
19	6 m 90 cm + 10 cm =	m	cm	41	5 m 85 cm + 3 m 45 cm =	m	cm
20	6 m 90 cm + 20 cm =	m	cm	42	4 m 87 cm + 3 m 76 cm =	m	cm
21	6 m 90 cm + 30 cm =	m	cm	43	6 m 36 cm + 4 m 67 cm =	m	cm
22	6 m 90 cm + 60 cm =	m	cm	44	9 m 74 cm + 8 m 48 cm =	m	cm

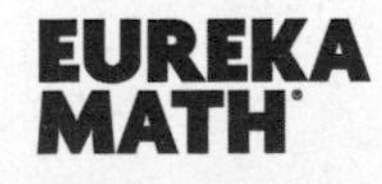

A

Number Correct: ________

Convert to Kilograms and Grams

1.	2,000 g =	2 kg 000 g
2.	3,000 g =	3 kg 000 g
3.	4,000 g =	4 kg 000 g
4.	9,000 g =	9 kg 000 g
5.	6,000 g =	6 kg 000 g
6.	1,000 g =	1 kg 000 g
7.	8,000 g =	8 kg 000 g
8.	5,000 g =	5 kg 000 g
9.	7,000 g =	7 kg 000 g
10.	6,100 g =	6 kg 100 g
11.	6,110 g =	6 kg 110 g
12.	6,101 g =	6 kg 101 g
13.	6,010 g =	6 kg 010 g
14.	6,011 g =	6 kg 011 g
15.	6,001 g =	6 kg 001 g
16.	8,002 g =	8 kg 002 g
17.	8,020 g =	8 kg 020 g
18.	8,200 g =	8 kg 200 g
19.	8,022 g =	8 kg 022 g
20.	8,220 g =	8 kg 220 g
21.	8,222 g =	8 kg 222 g
22.	7,256 g =	7 kg 256 g
23.	3,800 g =	3 kg 800 g
24.	4,770 g =	4 kg 770 g
25.	4,807 g =	4 kg 807 g
26.	5,065 g =	5 kg 065 g
27.	5,040 g =	5 kg 040 g
28.	6,007 g =	6 kg 007 g
29.	2,003 g =	2 kg 003 g
30.	1,090 g =	1 kg 090 g
31.	1,055 g =	1 kg 055 g
32.	9,404 g =	9 kg 404 g
33.	9,330 g =	9 kg 330 g
34.	3,400 g =	3 kg 400 g
35.	4,000 g + 2,000 g =	6 kg 000 g
36.	5,000 g + 3,000 g =	8 kg 000 g
37.	4,000 g + 4,000 g =	8 kg 000 g
38.	8 × 7,000 g =	56 kg 000 g
39.	49,000 g ÷ 7 =	kg g
40.	16,000 g × 5 =	kg g
41.	63,000 g ÷ 7 =	kg g
42.	17 × 4,000 g =	kg g
43.	13,000 g × 5 =	kg g
44.	84,000 g ÷ 7 =	kg g

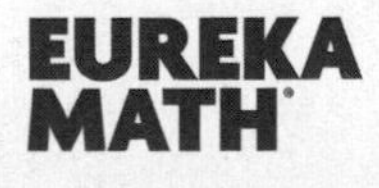

B

Number Correct: ________

Convert to Kilograms and Grams

Improvement: ________

1.	1,000 g =	kg	g
2.	2,000 g =	kg	g
3.	3,000 g =	kg	g
4.	8,000 g =	kg	g
5.	6,000 g =	kg	g
6.	9,000 g =	kg	g
7.	4,000 g =	kg	g
8.	7,000 g =	kg	g
9.	5,000 g =	kg	g
10.	5,100 g =	kg	g
11.	5,110 g =	kg	g
12.	5,101 g =	kg	g
13.	5,010 g =	kg	g
14.	5,011 g =	kg	g
15.	5,001 g =	kg	g
16.	7,002 g =	kg	g
17.	7,020 g =	kg	g
18.	7,200 g =	kg	g
19.	7,022 g =	kg	g
20.	7,220 g =	kg	g
21.	7,222 g =	kg	g
22.	4,378 g =	kg	g

23.	2,700 g =	kg	g
24.	3,660 g =	kg	g
25.	3,706 g =	kg	g
26.	4,095 g =	kg	g
27.	4,030 g =	kg	g
28.	5,006 g =	kg	g
29.	3,004 g =	kg	g
30.	2,010 g =	kg	g
31.	2,075 g =	kg	g
32.	1,504 g =	kg	g
33.	1,440 g =	kg	g
34.	4,500 g =	kg	g
35.	3,000 g + 2,000 g =	kg	g
36.	4,000 g + 3,000 g =	kg	g
37.	5,000 g + 4,000 g =	kg	g
38.	9 × 8,000 g =	kg	g
39.	64,000 g ÷ 8 =	kg	g
40.	17,000 g × 5 =	kg	g
41.	54,000 g ÷ 6 =	kg	g
42.	18,000 g × 4 =	kg	g
43.	14 × 5,000 g =	kg	g
44.	96,000 g ÷ 8 =	kg	g

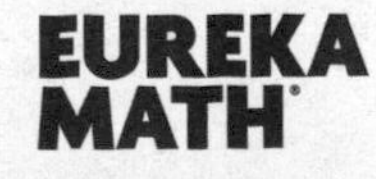

Grade 4
Module 3

A

Number Correct: ______

Squares and Unknown Factors

1.	2 × 2 =	4
2.	2 × __2__ = 4	
3.	3 × 3 =	9
4.	3 × __3__ = 9	
5.	5 × 5 =	25
6.	5 × __5__ = 25	
7.	1 × __1__ = 1	1
8.	1 × 1 =	1
9.	4 × __4__ = 16	16
10.	4 × 4 =	16
11.	7 × __7__ = 49	49
12.	7 × 7 =	49
13.	8 × 8 =	64
14.	8 × __8__ = 64	
15.	10 × 10 =	100
16.	10 × __10__ = 100	
17.	9 × __9__ = 81	
18.	9 × 9 =	81
19.	2 × __5__ = 10	
20.	2 × ______ = 18	
21.	2 × 2 =	
22.	3 × ______ = 12	
23.	3 × ______ = 21	
24.	3 × 3 =	
25.	4 × ______ = 20	
26.	4 × ______ = 32	
27.	4 × 4 =	
28.	5 × ______ = 20	
29.	5 × ______ = 40	
30.	5 × 5 =	
31.	6 × ______ = 18	
32.	6 × ______ = 54	
33.	6 × 6 =	
34.	7 × ______ = 28	
35.	7 × ______ = 56	
36.	7 × 7 =	
37.	8 × ______ = 24	
38.	8 × ______ = 72	
39.	8 × 8 =	
40.	9 × ______ = 36	
41.	9 × ______ = 63	
42.	9 × 9 =	
43.	9 × ______ = 54	
44.	10 × 10 =	

B

Number Correct: ______

Squares and Unknown Factors

Improvement: ______

1.	5 × 5 =	25
2.	5 × 5 = 25	
3.	2 × 2 =	4
4.	2 × 2 = 4	
5.	3 × 3 =	9
6.	3 × 3 = 9	
7.	1 × 1 =	1
8.	1 × 1 = 1	
9.	4 × 4 = 16	
10.	4 × 4 =	16
11.	6 × 6 = 36	
12.	6 × 6 =	36
13.	9 × 9 =	81
14.	9 × 9 = 81	
15.	10 × 10 =	100
16.	10 × 10 = 100	
17.	7 × 7 = 49	
18.	7 × 7 =	49
19.	2 × 4 = 8	
20.	2 × 8 = 16	
21.	2 × 2 =	4
22.	3 × 5 = 15	
23.	3 × 8 = 24	
24.	3 × 3 =	9
25.	4 × 3 = 12	
26.	4 × 7 = 28	
27.	4 × 4 =	16
28.	5 × 2 = 10	
29.	5 × 7 = 35	
30.	5 × 5 =	25
31.	6 × ____ = 24	
32.	6 × ____ = 48	
33.	6 × 6 =	
34.	7 × ____ = 21	
35.	7 × ____ = 63	
36.	7 × 7 =	
37.	8 × ____ = 32	
38.	8 × ____ = 56	
39.	8 × 8 =	
40.	9 × ____ = 27	
41.	9 × ____ = 72	
42.	9 × 9 =	
43.	9 × ____ = 63	
44.	10 × 10 =	

A

Number Correct: ______

Multiply Multiples of 10, 100, and 1,000

1.	3 × 2 =	6
2.	30 × 2 =	60
3.	300 × 2 =	600
4.	3,000 × 2 =	6,000
5.	2 × 3,000 =	6,000
6.	2 × 4 =	8
7.	2 × 40 =	80
8.	2 × 400 =	800
9.	2 × 4,000 =	8,000
10.	3 × 3 =	9
11.	30 × 3 =	90
12.	300 × 3 =	900
13.	3,000 × 3 =	9000
14.	4,000 × 3 =	12,000
15.	400 × 3 =	1
16.	40 × 3 =	
17.	5 × 3 =	
18.	500 × 3 =	
19.	7 × 2 =	
20.	70 × 2 =	
21.	4 × 4 =	
22.	4,000 × 4 =	

23.	7 × 5 =	
24.	700 × 5 =	
25.	8 × 3 =	
26.	80 × 3 =	
27.	9 × 4 =	
28.	9,000 × 4 =	
29.	7 × 6 =	
30.	7 × 600 =	
31.	8 × 9 =	
32.	8 × 90 =	
33.	6 × 9 =	
34.	6 × 9,000 =	
35.	900 × 9 =	
36.	8,000 × 8 =	
37.	7 × 70 =	
38.	6 × 600 =	
39.	800 × 7 =	
40.	7 × 9,000 =	
41.	200 × 5 =	
42.	5 × 60 =	
43.	4,000 × 5 =	
44.	800 × 5 =	

B

Number Correct: ________

Improvement: ________

Multiply Multiples of 10, 100, and 1,000

1.	4 × 2 =	8
2.	40 × 2 =	80
3.	400 × 2 =	800
4.	4,000 × 2 =	8,000
5.	2 × 4,000 =	8,000
6.	3 × 3 =	9
7.	3 × 30 =	90
8.	3 × 300 =	900
9.	3 × 3,000 =	9,000
10.	2 × 3 =	6
11.	20 × 3 =	60
12.	200 × 3 =	600
13.	2,000 × 3 =	6,060
14.	3,000 × 4 =	12,000
15.	300 × 4 =	1,200
16.	30 × 4 =	
17.	3 × 5 =	
18.	30 × 5 =	
19.	6 × 2 =	
20.	60 × 2 =	
21.	4 × 4 =	
22.	400 × 4 =	

23.	9 × 5 =	
24.	900 × 5 =	
25.	8 × 4 =	
26.	80 × 4 =	
27.	9 × 3 =	
28.	9,000 × 3 =	
29.	6 × 7 =	
30.	6 × 700 =	
31.	8 × 7 =	
32.	8 × 70 =	
33.	9 × 6 =	
34.	9 × 6,000 =	
35.	800 × 8 =	
36.	9,000 × 9 =	
37.	7 × 700 =	
38.	6 × 60 =	
39.	700 × 8 =	
40.	9 × 7,000 =	
41.	20 × 5 =	
42.	5 × 600 =	
43.	400 × 5 =	
44.	8,000 × 5 =	

A

Number Correct: _______

Mental Multiplication

1.	1 × 4 =	4
2.	10 × 4 =	40
3.	11 × 4 =	44
4.	1 × 2 =	2
5.	20 × 2 =	40
6.	21 × 2 =	42
7.	2 × 3 =	6
8.	30 × 3 =	90
9.	32 × 3 =	96
10.	3 × 5 =	15
11.	20 × 5 =	100
12.	23 × 5 =	
13.	3 × 3 =	
14.	40 × 3 =	
15.	43 × 3 =	
16.	4 × 2 =	
17.	70 × 2 =	
18.	74 × 2 =	
19.	2 × 3 =	
20.	60 × 3 =	
21.	62 × 3 =	
22.	63 × 3 =	
23.	21 × 3 =	
24.	121 × 3 =	
25.	42 × 2 =	
26.	142 × 2 =	
27.	242 × 2 =	
28.	342 × 2 =	
29.	442 × 2 =	
30.	3 × 3 =	
31.	13 × 3 =	
32.	213 × 3 =	
33.	1,213 × 3 =	
34.	2,113 × 3 =	
35.	2,131 × 3 =	
36.	2,311 × 3 =	
37.	24 × 4 =	
38.	35 × 5 =	
39.	54 × 3 =	
40.	63 × 6 =	
41.	125 × 4 =	
42.	214 × 3 =	
43.	5,213 × 2 =	
44.	2,135 × 4 =	

B

Number Correct: ______

Mental Multiplication

Improvement: ______

1.	1 × 6 =	6
2.	10 × 6 =	60
3.	11 × 6 =	66
4.	1 × 2 =	
5.	30 × 2 =	60
6.	31 × 2 =	62
7.	3 × 3 =	9
8.	20 × 3 =	60
9.	23 × 3 =	69
10.	5 × 5 =	25
11.	20 × 5 =	100
12.	25 × 5 =	125
13.	4 × 4 =	16
14.	30 × 4 =	
15.	34 × 4 =	
16.	4 × 2 =	
17.	90 × 2 =	
18.	94 × 2 =	
19.	2 × 3 =	
20.	40 × 3 =	
21.	42 × 3 =	
22.	43 × 3 =	

23.	21 × 4 =	
24.	121 × 4 =	
25.	24 × 2 =	
26.	124 × 2 =	
27.	224 × 2 =	
28.	324 × 2 =	
29.	424 × 2 =	
30.	3 × 2 =	
31.	13 × 2 =	
32.	213 × 2 =	
33.	1,213 × 2 =	
34.	2,113 × 2 =	
35.	2,131 × 2 =	
36.	2,311 × 2 =	
37.	23 × 4 =	
38.	53 × 5 =	
39.	45 × 3 =	
40.	36 × 6 =	
41.	215 × 3 =	
42.	125 × 4 =	
43.	5,312 × 2 =	
44.	1,235 × 4 =	

A

Number Correct: _______

Mental Division

1.	20 ÷ 2 =		23.	68 ÷ 2 =	
2.	4 ÷ 2 =		24.	96 ÷ 3 =	
3.	24 ÷ 2 =		25.	86 ÷ 2 =	
4.	30 ÷ 3 =		26.	93 ÷ 3 =	
5.	6 ÷ 3 =		27.	88 ÷ 4 =	
6.	36 ÷ 3 =		28.	99 ÷ 3 =	
7.	40 ÷ 4 =		29.	66 ÷ 3 =	
8.	8 ÷ 4 =		30.	66 ÷ 2 =	
9.	48 ÷ 4 =		31.	40 ÷ 4 =	
10.	2 ÷ 2 =		32.	80 ÷ 4 =	
11.	40 ÷ 2 =		33.	60 ÷ 4 =	
12.	42 ÷ 2 =		34.	68 ÷ 4 =	
13.	3 ÷ 3 =		35.	20 ÷ 2 =	
14.	60 ÷ 3 =		36.	40 ÷ 2 =	
15.	63 ÷ 3 =		37.	30 ÷ 2 =	
16.	4 ÷ 4 =		38.	36 ÷ 2 =	
17.	80 ÷ 4 =		39.	30 ÷ 3 =	
18.	84 ÷ 4 =		40.	39 ÷ 3 =	
19.	40 ÷ 5 =		41.	45 ÷ 3 =	
20.	50 ÷ 5 =		42.	60 ÷ 3 =	
21.	60 ÷ 5 =		43.	57 ÷ 3 =	
22.	70 ÷ 5 =		44.	51 ÷ 3 =	

B

Number Correct: _______

Mental Division

Improvement: _______

1.	30 ÷ 3 =	
2.	9 ÷ 3 =	
3.	39 ÷ 3 =	
4.	20 ÷ 2 =	
5.	6 ÷ 2 =	
6.	26 ÷ 2 =	
7.	80 ÷ 4 =	
8.	4 ÷ 4 =	
9.	84 ÷ 4 =	
10.	2 ÷ 2 =	
11.	60 ÷ 2 =	
12.	62 ÷ 2 =	
13.	3 ÷ 3 =	
14.	90 ÷ 3 =	
15.	93 ÷ 3 =	
16.	8 ÷ 4 =	
17.	40 ÷ 4 =	
18.	48 ÷ 4 =	
19.	50 ÷ 5 =	
20.	60 ÷ 5 =	
21.	70 ÷ 5 =	
22.	80 ÷ 5 =	

23.	86 ÷ 2 =	
24.	69 ÷ 3 =	
25.	68 ÷ 2 =	
26.	96 ÷ 3 =	
27.	66 ÷ 3 =	
28.	99 ÷ 3 =	
29.	88 ÷ 4 =	
30.	88 ÷ 2 =	
31.	40 ÷ 4 =	
32.	80 ÷ 4 =	
33.	60 ÷ 4 =	
34.	64 ÷ 4 =	
35.	20 ÷ 2 =	
36.	40 ÷ 2 =	
37.	30 ÷ 2 =	
38.	38 ÷ 2 =	
39.	30 ÷ 3 =	
40.	36 ÷ 3 =	
41.	42 ÷ 3 =	
42.	60 ÷ 3 =	
43.	54 ÷ 3 =	
44.	48 ÷ 3 =	

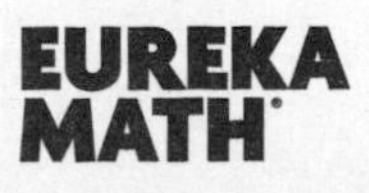

A

Number Correct: ______

Division with Remainders

1.	8 ÷ 2	Q = ______	R = ______	23.	6 ÷ 2	Q = ______	R = ______
2.	9 ÷ 2	Q = ______	R = ______	24.	7 ÷ 2	Q = ______	R = ______
3.	4 ÷ 4	Q = ______	R = ______	25.	3 ÷ 3	Q = ______	R = ______
4.	5 ÷ 4	Q = ______	R = ______	26.	4 ÷ 3	Q = ______	R = ______
5.	7 ÷ 5	Q = ______	R = ______	27.	6 ÷ 4	Q = ______	R = ______
6.	8 ÷ 5	Q = ______	R = ______	28.	7 ÷ 4	Q = ______	R = ______
7.	5 ÷ 3	Q = ______	R = ______	29.	6 ÷ 6	Q = ______	R = ______
8.	6 ÷ 3	Q = ______	R = ______	30.	7 ÷ 6	Q = ______	R = ______
9.	8 ÷ 4	Q = ______	R = ______	31.	4 ÷ 2	Q = ______	R = ______
10.	9 ÷ 4	Q = ______	R = ______	32.	5 ÷ 2	Q = ______	R = ______
11.	2 ÷ 2	Q = ______	R = ______	33.	9 ÷ 3	Q = ______	R = ______
12.	3 ÷ 2	Q = ______	R = ______	34.	9 ÷ 5	Q = ______	R = ______
13.	7 ÷ 3	Q = ______	R = ______	35.	7 ÷ 7	Q = ______	R = ______
14.	8 ÷ 3	Q = ______	R = ______	36.	9 ÷ 9	Q = ______	R = ______
15.	9 ÷ 3	Q = ______	R = ______	37.	13 ÷ 4	Q = ______	R = ______
16.	8 ÷ 6	Q = ______	R = ______	38.	18 ÷ 5	Q = ______	R = ______
17.	9 ÷ 6	Q = ______	R = ______	39.	21 ÷ 6	Q = ______	R = ______
18.	5 ÷ 5	Q = ______	R = ______	40.	24 ÷ 7	Q = ______	R = ______
19.	6 ÷ 5	Q = ______	R = ______	41.	29 ÷ 8	Q = ______	R = ______
20.	8 ÷ 8	Q = ______	R = ______	42.	43 ÷ 6	Q = ______	R = ______
21.	9 ÷ 8	Q = ______	R = ______	43.	53 ÷ 6	Q = ______	R = ______
22.	9 ÷ 9	Q = ______	R = ______	44.	82 ÷ 9	Q = ______	R = ______

B

Number Correct: ______

Division with Remainders

Improvement: ______

1.	9 ÷ 8	Q = 8	R = 1
2.	8 ÷ 8	Q = 6	R = 1
3.	9 ÷ 6	Q = ______	R = ______
4.	8 ÷ 6	Q = ______	R = ______
5.	5 ÷ 5	Q = ______	R = ______
6.	6 ÷ 5	Q = ______	R = ______
7.	7 ÷ 4	Q = ______	R = ______
8.	6 ÷ 4	Q = ______	R = ______
9.	5 ÷ 3	Q = ______	R = ______
10.	6 ÷ 3	Q = 2	R = ______
11.	2 ÷ 2	Q = ______	R = ______
12.	3 ÷ 2	Q = ______	R = ______
13.	3 ÷ 3	Q = ______	R = ______
14.	4 ÷ 3	Q = ______	R = ______
15.	8 ÷ 7	Q = ______	R = ______
16.	9 ÷ 7	Q = ______	R = ______
17.	4 ÷ 4	Q = ______	R = ______
18.	5 ÷ 4	Q = ______	R = ______
19.	6 ÷ 2	Q = ______	R = ______
20.	7 ÷ 2	Q = ______	R = ______
21.	8 ÷ 5	Q = ______	R = ______
22.	7 ÷ 5	Q = ______	R = ______
23.	4 ÷ 2	Q = ______	R = ______
24.	5 ÷ 2	Q = ______	R = ______
25.	8 ÷ 4	Q = ______	R = ______
26.	9 ÷ 4	Q = ______	R = ______
27.	9 ÷ 3	Q = ______	R = ______
28.	8 ÷ 3	Q = ______	R = ______
29.	9 ÷ 5	Q = ______	R = ______
30.	6 ÷ 6	Q = 1	R = 1
31.	7 ÷ 6	Q = ______	R = ______
32.	9 ÷ 9	Q = ______	R = ______
33.	7 ÷ 7	Q = ______	R = ______
34.	9 ÷ 2	Q = ______	R = ______
35.	8 ÷ 2	Q = ______	R = ______
36.	37 ÷ 8	Q = ______	R = ______
37.	50 ÷ 9	Q = ______	R = ______
38.	17 ÷ 6	Q = ______	R = ______
39.	48 ÷ 7	Q = ______	R = ______
40.	51 ÷ 8	Q = ______	R = ______
41.	68 ÷ 9	Q = ______	R = ______
42.	53 ÷ 6	Q = ______	R = ______
43.	61 ÷ 8	Q = ______	R = ______
44.	70 ÷ 9	Q = ______	R = ______

A

Number Correct: ______

Circle the Prime Number

1.	4	3
2.	6	3
3.	8	3
4.	5	10
5.	5	12
6.	5	14
7.	8	7
8.	9	11
9.	11	15
10.	15	17
11.	19	16
12.	14	11
13.	13	12
14.	18	17
15.	19	20
16.	21	23
17.	25	19
18.	29	27
19.	31	30
20.	33	37
21.	9	2
22.	51	2

23.	40	41	42
24.	42	43	44
25.	49	47	45
26.	53	50	55
27.	54	56	59
28.	99	97	95
29.	89	90	91
30.	95	96	97
31.	88	89	90
32.	60	61	62
33.	63	65	67
34.	71	70	69
35.	73	75	77
36.	49	79	99
37.	63	93	83
38.	22	2	12
39.	17	27	57
40.	5	15	25
41.	39	49	59
42.	1	21	31
43.	51	57	2
44.	84	95	43

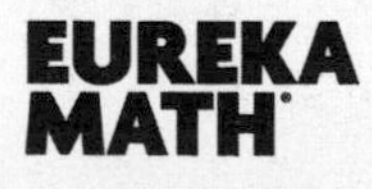

B

Number Correct: ______

Circle the Prime Number

Improvement: ______

1.	4	5
2.	6	5
3.	8	5
4.	7	10
5.	7	12
6.	7	14
7.	4	3
8.	11	10
9.	15	11
10.	17	15
11.	19	20
12.	14	13
13.	11	12
14.	16	17
15.	19	18
16.	22	23
17.	21	19
18.	29	28
19.	31	33
20.	35	37
21.	2	9
22.	57	2

23.	42	41	40
24.	44	43	42
25.	45	47	49
26.	53	55	50
27.	56	54	59
28.	95	97	99
29.	91	90	89
30.	99	98	97
31.	90	89	88
32.	67	65	63
33.	62	61	60
34.	72	71	70
35.	77	75	73
36.	27	67	77
37.	39	49	59
38.	32	2	22
39.	19	49	69
40.	5	15	55
41.	99	49	59
42.	1	21	41
43.	45	51	2
44.	48	85	67

A

Number Correct: _______

Divide.

1.	$6 \div 2 =$	
2.	$60 \div 2 =$	
3.	$600 \div 2 =$	
4.	$6,000 \div 2 =$	
5.	$9 \div 3 =$	
6.	$90 \div 3 =$	
7.	$900 \div 3 =$	
8.	$9,000 \div 3 =$	
9.	$10 \div 5 =$	
10.	$15 \div 5 =$	
11.	$150 \div 5 =$	
12.	$1,500 \div 5 =$	
13.	$2,500 \div 5 =$	
14.	$3,500 \div 5 =$	
15.	$4,500 \div 5 =$	
16.	$450 \div 5 =$	
17.	$8 \div 4 =$	
18.	$12 \div 4 =$	
19.	$120 \div 4 =$	
20.	$1,200 \div 4 =$	
21.	$25 \div 5 =$	
22.	$30 \div 5 =$	

23.	$300 \div 5 =$	
24.	$3,000 \div 5 =$	
25.	$16 \div 4 =$	
26.	$160 \div 4 =$	
27.	$18 \div 6 =$	
28.	$1,800 \div 6 =$	
29.	$28 \div 7 =$	
30.	$280 \div 7 =$	
31.	$48 \div 8 =$	
32.	$4,800 \div 8 =$	
33.	$6,300 \div 9 =$	
34.	$200 \div 5 =$	
35.	$560 \div 7 =$	
36.	$7,200 \div 9 =$	
37.	$480 \div 6 =$	
38.	$5,600 \div 8 =$	
39.	$400 \div 5 =$	
40.	$6,300 \div 7 =$	
41.	$810 \div 9 =$	
42.	$640 \div 8 =$	
43.	$5,400 \div 6 =$	
44.	$4,000 \div 5 =$	

B

Number Correct: ________

Divide.

Improvement: ________

1.	4 ÷ 2 =		23.	200 ÷ 5 =	
2.	40 ÷ 2 =		24.	2,000 ÷ 5 =	
3.	400 ÷ 2 =		25.	12 ÷ 4 =	
4.	4,000 ÷ 2 =		26.	120 ÷ 4 =	
5.	6 ÷ 3 =		27.	21 ÷ 7 =	
6.	60 ÷ 3 =		28.	2,100 ÷ 7 =	
7.	600 ÷ 3 =		29.	18 ÷ 6 =	
8.	6,000 ÷ 3 =		30.	180 ÷ 6 =	
9.	10 ÷ 5 =		31.	54 ÷ 9 =	
10.	15 ÷ 5 =		32.	5,400 ÷ 9 =	
11.	150 ÷ 5 =		33.	5,600 ÷ 8 =	
12.	250 ÷ 5 =		34.	300 ÷ 5 =	
13.	350 ÷ 5 =		35.	490 ÷ 7 =	
14.	3,500 ÷ 5 =		36.	6,300 ÷ 9 =	
15.	4,500 ÷ 5 =		37.	420 ÷ 6 =	
16.	450 ÷ 5 =		38.	4,800 ÷ 8 =	
17.	9 ÷ 3 =		39.	4,000 ÷ 5 =	
18.	12 ÷ 3 =		40.	560 ÷ 8 =	
19.	120 ÷ 3 =		41.	6,400 ÷ 8 =	
20.	1,200 ÷ 3 =		42.	720 ÷ 8 =	
21.	25 ÷ 5 =		43.	4,800 ÷ 6 =	
22.	20 ÷ 5 =		44.	400 ÷ 5 =	

Grade 4
Module 5

Note: Module 4 does not contain any Sprints or printed fluency components. All fluency activities in Module 4 can be found in the Teacher Edition and can be completed without paper.

A

Number Correct: ______

Multiply Whole Numbers Times Fractions

1.	$\frac{1}{3}+\frac{1}{3}=$	
2.	$2\times\frac{1}{3}=$	
3.	$\frac{1}{4}+\frac{1}{4}+\frac{1}{4}=$	
4.	$3\times\frac{1}{4}=$	
5.	$\frac{1}{5}+\frac{1}{5}=$	
6.	$2\times\frac{1}{5}=$	
7.	$\frac{1}{5}+\frac{1}{5}+\frac{1}{5}=$	
8.	$3\times\frac{1}{5}=$	
9.	$\frac{1}{5}+\frac{1}{5}+\frac{1}{5}+\frac{1}{5}=$	
10.	$4\times\frac{1}{5}=$	
11.	$\frac{1}{10}+\frac{1}{10}+\frac{1}{10}=$	
12.	$3\times\frac{1}{10}=$	
13.	$\frac{1}{8}+\frac{1}{8}+\frac{1}{8}=$	
14.	$3\times\frac{1}{8}=$	
15.	$\frac{1}{2}+\frac{1}{2}=$	
16.	$2\times\frac{1}{2}=$	
17.	$\frac{1}{3}+\frac{1}{3}+\frac{1}{3}=$	
18.	$3\times\frac{1}{3}=$	
19.	$\frac{1}{4}+\frac{1}{4}+\frac{1}{4}+\frac{1}{4}=$	
20.	$4\times\frac{1}{4}=$	
21.	$\frac{1}{2}+\frac{1}{2}+\frac{1}{2}=$	
22.	$3\times\frac{1}{2}=$	

23.	$\frac{1}{3}+\frac{1}{3}+\frac{1}{3}+\frac{1}{3}=$	
24.	$4\times\frac{1}{3}=$	
25.	$\frac{5}{6}=$	$\_\_\times\frac{1}{6}$
26.	$\frac{5}{6}=$	$5\times$ —
27.	$\frac{5}{8}=$	$5\times$ —
28.	$\frac{5}{8}=$	$\_\_\times\frac{1}{8}$
29.	$\frac{7}{8}=$	$7\times$ —
30.	$\frac{7}{10}=$	$7\times$ —
31.	$\frac{7}{8}=$	$\_\_\times\frac{1}{8}$
32.	$\frac{7}{10}=$	$\_\_\times\frac{1}{10}$
33.	$\frac{6}{6}=$	$6\times$ —
34.	$1=$	$6\times$ —
35.	$\frac{8}{8}=$	$\_\_\times\frac{1}{8}$
36.	$1=$	$\_\_\times\frac{1}{8}$
37.	$9\times\frac{1}{10}=$	
38.	$7\times\frac{1}{5}=$	
39.	$1=$	$3\times$ —
40.	$7\times\frac{1}{12}=$	
41.	$1=$	$\_\_\times\frac{1}{5}$
42.	$\frac{3}{5}=$	$\frac{1}{5}+\frac{1}{5}+$ –
43.	$3\times\frac{1}{4}=$	$\_\_+\frac{1}{4}+\frac{1}{4}$
44.	$1=$	— + — + —

B

Number Correct: ______

Multiply Whole Numbers Times Fractions

Improvement: ______

1.	$\frac{1}{5} + \frac{1}{5} =$	
2.	$2 \times \frac{1}{5} =$	
3.	$\frac{1}{3} + \frac{1}{3} =$	
4.	$2 \times \frac{1}{3} =$	
5.	$\frac{1}{4} + \frac{1}{4} + \frac{1}{4} =$	
6.	$3 \times \frac{1}{4} =$	
7.	$\frac{1}{5} + \frac{1}{5} + \frac{1}{5} =$	
8.	$3 \times \frac{1}{5} =$	
9.	$\frac{1}{5} + \frac{1}{5} + \frac{1}{5} + \frac{1}{5} =$	
10.	$4 \times \frac{1}{5} =$	
11.	$\frac{1}{8} + \frac{1}{8} + \frac{1}{8} =$	
12.	$3 \times \frac{1}{8} =$	
13.	$\frac{1}{10} + \frac{1}{10} + \frac{1}{10} =$	
14.	$3 \times \frac{1}{10} =$	
15.	$\frac{1}{3} + \frac{1}{3} + \frac{1}{3} =$	
16.	$3 \times \frac{1}{3} =$	
17.	$\frac{1}{4} + \frac{1}{4} + \frac{1}{4} + \frac{1}{4} =$	
18.	$4 \times \frac{1}{4} =$	
19.	$\frac{1}{2} + \frac{1}{2} =$	
20.	$2 \times \frac{1}{2} =$	
21.	$\frac{1}{3} + \frac{1}{3} + \frac{1}{3} + \frac{1}{3} =$	
22.	$4 \times \frac{1}{3} =$	
23.	$\frac{1}{2} + \frac{1}{2} + \frac{1}{2} =$	
24.	$3 \times \frac{1}{2} =$	
25.	$\frac{5}{6} =$	$\_\_ \times \frac{1}{6}$
26.	$\frac{5}{6} =$	$5 \times \_\_$
27.	$\frac{5}{8} =$	$5 \times \_\_$
28.	$\frac{5}{8} =$	$\_\_ \times \frac{1}{8}$
29.	$\frac{7}{8} =$	$7 \times \_\_$
30.	$\frac{7}{10} =$	$7 \times \_\_$
31.	$\frac{7}{8} =$	$\_\_ \times \frac{1}{8}$
32.	$\frac{7}{10} =$	$\_\_ \times \frac{1}{10}$
33.	$\frac{8}{8} =$	$8 \times \_\_$
34.	$1 =$	$8 \times \_\_$
35.	$\frac{6}{6} =$	$\_\_ \times \frac{1}{6}$
36.	$1 =$	$\_\_ \times \frac{1}{6}$
37.	$5 \times \frac{1}{12} =$	
38.	$6 \times \frac{1}{5} =$	
39.	$1 =$	$4 \times \_\_$
40.	$9 \times \frac{1}{10} =$	
41.	$1 =$	$\_\_ \times \frac{1}{3}$
42.	$\frac{3}{4} =$	$\frac{1}{4} + \frac{1}{4} + \_\_$
43.	$3 \times \frac{1}{5} =$	$\_ + \frac{1}{5} + \frac{1}{5}$
44.	$1 =$	$\_\_ + \_\_ + \_\_ + \_\_$

A

Number Correct: ________

Subtract Fractions

1.	$2 - 1 =$	
2.	$\frac{2}{2} - \frac{1}{2} =$	
3.	$1 - \frac{1}{2} =$	
4.	$3 - 1 =$	
5.	$\frac{3}{3} - \frac{1}{3} =$	
6.	$1 - \frac{1}{3} =$	
7.	$8 - 1 =$	
8.	$\frac{8}{8} - \frac{1}{8} =$	
9.	$1 - \frac{1}{8} =$	
10.	$5 - 1 =$	
11.	$\frac{5}{5} - \frac{1}{5} =$	
12.	$1 - \frac{1}{5} =$	
13.	$1 - \frac{2}{5} =$	
14.	$1 - \frac{4}{5} =$	
15.	$1 - \frac{3}{5} =$	
16.	$1 - \frac{1}{4} =$	
17.	$1 - \frac{3}{4} =$	
18.	$1 - \frac{1}{10} =$	
19.	$1 - \frac{9}{10} =$	
20.	$1 - \frac{3}{10} =$	
21.	$1 - \frac{7}{10} =$	
22.	$4 - 2 =$	
23.	$\frac{4}{3} - \frac{2}{3} =$	
24.	$1\frac{1}{3} - \frac{2}{3} =$	
25.	$1\frac{2}{3} - \frac{1}{3} =$	
26.	$7 - 4 =$	
27.	$\frac{7}{5} - \frac{4}{5} =$	
28.	$1\frac{2}{5} - \frac{4}{5} =$	
29.	$1\frac{4}{5} - \frac{2}{5} =$	
30.	$5 - 3 =$	
31.	$\frac{5}{4} - \frac{3}{4} =$	
32.	$1\frac{1}{4} - \frac{3}{4} =$	
33.	$1\frac{3}{4} - \frac{1}{4} =$	
34.	$1 - \frac{3}{8} =$	
35.	$1 - \frac{7}{8} =$	
36.	$1\frac{7}{8} - \frac{3}{8} =$	
37.	$1\frac{3}{8} - \frac{7}{8} =$	
38.	$1 - \frac{1}{6} =$	
39.	$1 - \frac{5}{6} =$	
40.	$1\frac{5}{6} - \frac{1}{6} =$	
41.	$1\frac{1}{6} - \frac{5}{6} =$	
42.	$1 - \frac{5}{12} =$	
43.	$1\frac{1}{12} - \frac{7}{12} =$	
44.	$1\frac{4}{15} - \frac{13}{15} =$	

B

Number Correct: _______

Subtract Fractions

Improvement: _______

1.	$3 - 1 =$		23.	$\frac{5}{4} - \frac{3}{4} =$	
2.	$\frac{3}{3} - \frac{1}{3} =$		24.	$1\frac{1}{4} - \frac{3}{4} =$	
3.	$1 - \frac{1}{3} =$		25.	$1\frac{3}{4} - \frac{1}{4} =$	
4.	$2 - 1 =$		26.	$8 - 4 =$	
5.	$\frac{2}{2} - \frac{1}{2} =$		27.	$\frac{8}{5} - \frac{4}{5} =$	
6.	$1 - \frac{1}{2} =$		28.	$1\frac{3}{5} - \frac{4}{5} =$	
7.	$6 - 1 =$		29.	$1\frac{4}{5} - \frac{3}{5} =$	
8.	$\frac{6}{6} - \frac{1}{6} =$		30.	$7 - 5 =$	
9.	$1 - \frac{1}{6} =$		31.	$\frac{7}{6} - \frac{5}{6} =$	
10.	$10 - 1 =$		32.	$1\frac{1}{6} - \frac{5}{6} =$	
11.	$\frac{10}{10} - \frac{1}{10} =$		33.	$1\frac{5}{6} - \frac{1}{6} =$	
12.	$1 - \frac{1}{10} =$		34.	$1 - \frac{5}{8} =$	
13.	$1 - \frac{2}{10} =$		35.	$1 - \frac{7}{8} =$	
14.	$1 - \frac{4}{10} =$		36.	$1\frac{7}{8} - \frac{5}{8} =$	
15.	$1 - \frac{3}{10} =$		37.	$1\frac{5}{8} - \frac{7}{8} =$	
16.	$1 - \frac{1}{5} =$		38.	$1 - \frac{1}{4} =$	
17.	$1 - \frac{4}{5} =$		39.	$1 - \frac{3}{4} =$	
18.	$1 - \frac{1}{8} =$		40.	$1\frac{3}{4} - \frac{1}{4} =$	
19.	$1 - \frac{7}{8} =$		41.	$1\frac{1}{4} - \frac{3}{4} =$	
20.	$1 - \frac{3}{8} =$		42.	$1 - \frac{7}{12} =$	
21.	$1 - \frac{5}{8} =$		43.	$1\frac{1}{12} - \frac{5}{12} =$	
22.	$5 - 3 =$		44.	$1\frac{7}{15} - \frac{11}{15} =$	

A

Number Correct: ______

Add Fractions

1.	$1 + 1 =$	
2.	$\frac{1}{5} + \frac{1}{5} =$	
3.	$2 + 1 =$	
4.	$\frac{2}{5} + \frac{1}{5} =$	
5.	$2 + 2 =$	
6.	$\frac{2}{5} + \frac{2}{5} =$	
7.	$3 + 2 =$	
8.	$\frac{3}{5} + \frac{2}{5} =$	fifths
9.	$\frac{5}{5} =$	
10.	$\frac{3}{5} + \frac{2}{5} =$	
11.	$3 + 2 =$	
12.	$\frac{3}{8} + \frac{2}{8} =$	
13.	$3 + 2 + 2 =$	
14.	$\frac{3}{8} + \frac{2}{8} + \frac{2}{8} =$	
15.	$\frac{3}{8} + \frac{3}{8} + \frac{2}{8} =$	eighths
16.	$\frac{8}{8} =$	
17.	$\frac{3}{8} + \frac{3}{8} + \frac{2}{8} =$	
18.	$2 + 1 + 1 =$	
19.	$\frac{2}{3} + \frac{1}{3} + \frac{1}{3} =$	thirds
20.	$\frac{2}{3} + \frac{1}{3} + \frac{1}{3} =$	$1\frac{}{3}$
21.	$2 + 2 + 2 =$	
22.	$\frac{2}{5} + \frac{2}{5} + \frac{2}{5} =$	fifths

23.	$\frac{2}{5} + \frac{2}{5} + \frac{2}{5} =$	$1\frac{}{5}$
24.	$3 + 3 + 3 =$	
25.	$\frac{3}{8} + \frac{3}{8} + \frac{3}{8} =$	eighths
26.	$\frac{3}{8} + \frac{3}{8} + \frac{3}{8} =$	$1\frac{}{8}$
27.	$\frac{5}{8} + \frac{5}{8} + \frac{5}{8} =$	$1\frac{}{8}$
28.	$1 + 1 + 1 =$	
29.	$\frac{1}{2} + \frac{1}{2} + \frac{1}{2} =$	halves
30.	$\frac{1}{2} + \frac{1}{2} + \frac{1}{2} =$	$1\frac{}{2}$
31.	$4 + 4 + 4 =$	
32.	$\frac{4}{10} + \frac{4}{10} + \frac{4}{10} =$	tenths
33.	$\frac{4}{10} + \frac{4}{10} + \frac{4}{10} =$	$1\frac{}{10}$
34.	$\frac{6}{10} + \frac{6}{10} + \frac{6}{10} =$	$1\frac{}{10}$
35.	$2 + 2 + 2 =$	
36.	$\frac{2}{6} + \frac{2}{6} + \frac{2}{6} =$	sixths
37.	$\frac{2}{6} + \frac{2}{6} + \frac{2}{6} =$	
38.	$\frac{3}{6} + \frac{3}{6} + \frac{3}{6} =$	$1\frac{}{6}$
39.	$\frac{5}{12} + \frac{2}{12} + \frac{4}{12} =$	
40.	$\frac{4}{12} + \frac{4}{12} + \frac{4}{12} =$	
41.	$\frac{5}{12} + \frac{5}{12} + \frac{7}{12} =$	$1\frac{}{12}$
42.	$\frac{7}{12} + \frac{9}{12} + \frac{7}{12} =$	$1\frac{}{12}$
43.	$\frac{7}{15} + \frac{8}{15} + \frac{7}{15} =$	$1\frac{}{15}$
44.	$\frac{12}{15} + \frac{8}{15} + \frac{9}{15} =$	$1\frac{}{15}$

B

Number Correct: _______

Add Fractions

Improvement: _______

1.	$1 + 1 =$	
2.	$\frac{1}{6} + \frac{1}{6} =$	
3.	$3 + 1 =$	
4.	$\frac{3}{6} + \frac{1}{6} =$	
5.	$3 + 2 =$	
6.	$\frac{3}{6} + \frac{2}{6} =$	
7.	$4 + 2 =$	
8.	$\frac{4}{6} + \frac{2}{6} =$	sixths
9.	$\frac{6}{6} =$	
10.	$\frac{4}{6} + \frac{2}{6} =$	
11.	$5 + 2 =$	
12.	$\frac{5}{8} + \frac{2}{8} =$	
13.	$5 + 1 + 1 =$	
14.	$\frac{5}{8} + \frac{1}{8} + \frac{1}{8} =$	
15.	$\frac{5}{8} + \frac{2}{8} + \frac{1}{8} =$	eighths
16.	$\frac{8}{8} =$	
17.	$\frac{3}{8} + \frac{3}{8} + \frac{2}{8} =$	
18.	$1 + 1 + 2 =$	
19.	$\frac{1}{3} + \frac{1}{3} + \frac{2}{3} =$	thirds
20.	$\frac{1}{3} + \frac{1}{3} + \frac{2}{3} =$	$1\frac{}{3}$
21.	$3 + 3 + 3 =$	
22.	$\frac{3}{8} + \frac{3}{8} + \frac{3}{8} =$	eighths

23.	$\frac{3}{8} + \frac{3}{8} + \frac{3}{8} =$	$1\frac{}{8}$
24.	$1 + 1 + 1 =$	
25.	$\frac{1}{2} + \frac{1}{2} + \frac{1}{2} =$	halves
26.	$\frac{1}{2} + \frac{1}{2} + \frac{1}{2} =$	$1\frac{}{2}$
27.	$2 + 2 + 2 =$	
28.	$\frac{2}{5} + \frac{2}{5} + \frac{2}{5} =$	fifths
29.	$\frac{2}{5} + \frac{2}{5} + \frac{2}{5} =$	$1\frac{}{5}$
30.	$\frac{3}{5} + \frac{3}{5} + \frac{3}{5} =$	$1\frac{}{5}$
31.	$6 + 6 + 6 =$	
32.	$\frac{6}{10} + \frac{6}{10} + \frac{6}{10} =$	tenths
33.	$\frac{6}{10} + \frac{6}{10} + \frac{6}{10} =$	$1\frac{}{10}$
34.	$\frac{5}{10} + \frac{5}{10} + \frac{5}{10} =$	$1\frac{}{10}$
35.	$2 + 2 + 2 =$	
36.	$\frac{2}{6} + \frac{2}{6} + \frac{2}{6} =$	sixths
37.	$\frac{2}{6} + \frac{2}{6} + \frac{2}{6} =$	
38.	$\frac{3}{6} + \frac{3}{6} + \frac{3}{6} =$	$1\frac{}{6}$
39.	$\frac{5}{12} + \frac{3}{12} + \frac{3}{12} =$	
40.	$\frac{5}{12} + \frac{5}{12} + \frac{2}{12} =$	
41.	$\frac{6}{12} + \frac{5}{12} + \frac{6}{12} =$	$1\frac{}{12}$
42.	$\frac{8}{12} + \frac{10}{12} + \frac{5}{12} =$	$1\frac{}{12}$
43.	$\frac{7}{15} + \frac{7}{15} + \frac{8}{15} =$	$1\frac{}{15}$
44.	$\frac{13}{15} + \frac{9}{15} + \frac{7}{15} =$	$1\frac{}{15}$

A

Number Correct: ______

Change Fractions to Mixed Numbers

1.	$3 = 2 + \_\_$	
2.	$\frac{3}{2} = \frac{2}{2} + \frac{}{2}$	
3.	$\frac{3}{2} = 1 + \frac{}{2}$	
4.	$\frac{3}{2} = 1\frac{}{2}$	
5.	$5 = 4 + \_\_$	
6.	$\frac{5}{4} = \frac{4}{4} + \frac{}{4}$	
7.	$\frac{5}{4} = 1 + \frac{}{4}$	
8.	$\frac{5}{4} = 1\frac{}{4}$	
9.	$4 = \_\_ + 1$	
10.	$\frac{4}{3} = \frac{}{3} + \frac{1}{3}$	
11.	$\frac{4}{3} = 1 + \frac{}{3}$	
12.	$\frac{4}{3} = \_\_\ \frac{1}{3}$	
13.	$7 = \_\_ + 2$	
14.	$\frac{7}{5} = \frac{}{5} + \frac{2}{5}$	
15.	$\frac{7}{5} = 1 + \frac{}{5}$	
16.	$\frac{7}{5} = 1\frac{}{5}$	
17.	$\frac{8}{5} = 1\frac{}{5}$	
18.	$\frac{9}{5} = 1\frac{}{5}$	
19.	$\frac{6}{5} = 1\frac{}{5}$	
20.	$\frac{10}{5} =$	
21.	$\frac{}{5} = \frac{10}{5} + \frac{1}{5}$	
22.	$\frac{}{5} = 2 + \frac{1}{5}$	

23.	$\frac{6}{3} =$	
24.	$\frac{}{3} = \frac{6}{3} + \frac{2}{3}$	
25.	$\frac{8}{3} = \frac{6}{3} + \frac{}{3}$	
26.	$\frac{8}{3} = 2 + \frac{}{3}$	
27.	$\frac{8}{3} = 2\frac{}{3}$	
28.	$\frac{}{4} = \frac{8}{4} + \frac{1}{4}$	
29.	$\frac{}{4} = 2 + \frac{1}{4}$	
30.	$\frac{9}{4} = \_\_\ \frac{1}{4}$	
31.	$\frac{11}{4} = \_\_\ \frac{3}{4}$	
32.	$\frac{8}{3} = \frac{}{3} + \frac{2}{3}$	
33.	$\frac{8}{3} = \frac{6}{3} + \frac{}{3}$	
34.	$\frac{8}{3} = \_\_ + \frac{2}{3}$	
35.	$\frac{8}{3} = \_\_\ \frac{2}{3}$	
36.	$\frac{14}{5} = \frac{10}{5} + \frac{}{5}$	
37.	$\frac{14}{5} = \_\_ + \frac{4}{5}$	
38.	$\frac{14}{5} = 2\frac{}{5}$	
39.	$\frac{13}{5} = 2\frac{}{5}$	
40.	$\frac{9}{8} = 1 + \frac{}{8}$	
41.	$\frac{15}{8} = 1 + \frac{}{8}$	
42.	$\frac{17}{12} = \frac{}{12} + \frac{5}{12}$	
43.	$\frac{11}{8} = 1 + \frac{}{8}$	
44.	$\frac{17}{12} = 1 + \frac{}{12}$	

B

Number Correct: ______

Change Fractions to Mixed Numbers

Improvement: ______

1.	$6 = 5 + \underline{\quad}$		23.	$\frac{4}{2} =$	
2.	$\frac{6}{5} = \frac{5}{5} + \frac{\ }{5}$		24.	$\frac{\ }{2} = \frac{4}{2} + \frac{1}{2}$	
3.	$\frac{6}{5} = 1 + \frac{\ }{5}$		25.	$\frac{5}{2} = \frac{4}{2} + \frac{\ }{2}$	
4.	$\frac{6}{5} = 1\frac{\ }{5}$		26.	$\frac{5}{2} = 2 + \frac{\ }{2}$	
5.	$4 = 3 + \underline{\quad}$		27.	$\frac{5}{2} = 2\frac{\ }{2}$	
6.	$\frac{4}{3} = \frac{3}{3} + \frac{\ }{3}$		28.	$\frac{\ }{5} = \frac{10}{5} + \frac{1}{5}$	
7.	$\frac{4}{3} = 1 + \frac{\ }{3}$		29.	$\frac{\ }{5} = 2 + \frac{1}{5}$	
8.	$\frac{4}{3} = 1\frac{\ }{3}$		30.	$\frac{11}{5} = \underline{\quad}\frac{1}{5}$	
9.	$5 = \underline{\quad} + 1$		31.	$\frac{13}{5} = \underline{\quad}\frac{3}{5}$	
10.	$\frac{5}{4} = \frac{\ }{4} + \frac{1}{4}$		32.	$\frac{5}{3} = \frac{\ }{3} + \frac{1}{3}$	
11.	$\frac{5}{4} = 1 + \frac{\ }{4}$		33.	$\frac{5}{2} = \frac{4}{2} + \frac{\ }{2}$	
12.	$\frac{5}{4} = \underline{\quad}\frac{1}{4}$		34.	$\frac{5}{2} = \underline{\quad} + \frac{1}{2}$	
13.	$8 = \underline{\quad} + 3$		35.	$\frac{5}{2} = \underline{\quad}\frac{1}{2}$	
14.	$\frac{8}{5} = \frac{\ }{5} + \frac{3}{5}$		36.	$\frac{12}{5} = \frac{10}{5} + \frac{\ }{5}$	
15.	$\frac{8}{5} = 1 + \frac{\ }{5}$		37.	$\frac{12}{5} = \underline{\quad} + \frac{2}{5}$	
16.	$\frac{8}{5} = 1\frac{\ }{5}$		38.	$\frac{12}{5} = 2\frac{\ }{5}$	
17.	$\frac{9}{5} = 1\frac{\ }{5}$		39.	$\frac{14}{5} = 2\frac{\ }{5}$	
18.	$\frac{6}{5} = 1\frac{\ }{5}$		40.	$\frac{9}{8} = 1 + \frac{\ }{8}$	
19.	$\frac{7}{5} = 1\frac{\ }{5}$		41.	$\frac{11}{8} = 1 + \frac{\ }{8}$	
20.	$\frac{6}{3} =$		42.	$\frac{19}{12} = \frac{\ }{12} + \frac{7}{12}$	
21.	$\frac{\ }{3} = \frac{6}{3} + \frac{1}{3}$		43.	$\frac{15}{8} = 1 + \frac{\ }{8}$	
22.	$\frac{\ }{3} = 2 + \frac{1}{3}$		44.	$\frac{19}{12} = 1 + \frac{\ }{12}$	

A

Number Correct: _______

Change Fractions to Mixed Numbers

1.	$3 + 1 =$	
2.	$\frac{3}{3} + \frac{1}{3} + \frac{}{3}$	
3.	$1 + \frac{1}{3} = \frac{}{3}$	
4.	$1\frac{1}{3} = \frac{}{3}$	
5.	$5 + 1 =$	
6.	$\frac{5}{5} + \frac{1}{5} = \frac{}{5}$	
7.	$1 + \frac{1}{5} = \frac{}{5}$	
8.	$1\frac{1}{5} = \frac{}{5}$	
9.	$2 + 1 =$	
10.	$\frac{2}{2} + \frac{1}{2} = \frac{}{2}$	
11.	$1 + \frac{1}{2} = \frac{}{2}$	
12.	$1\frac{1}{2} = \frac{}{2}$	
13.	$\frac{4}{4} + \frac{1}{4} = \frac{}{4}$	
14.	$1 + \frac{1}{4} = \frac{}{4}$	
15.	$1\frac{1}{4} = \frac{}{4}$	
16.	$1\frac{3}{4} = \frac{}{4}$	
17.	$\frac{5}{5} + \frac{1}{5} = \frac{}{5}$	
18.	$1 + \frac{1}{5} = \frac{}{5}$	
19.	$1\frac{1}{5} = \frac{}{5}$	
20.	$1\frac{3}{5} = \frac{}{5}$	
21.	$\frac{8}{8} + \frac{3}{8} = \frac{}{8}$	
22.	$1 + \frac{3}{8} = \frac{}{8}$	
23.	$1\frac{3}{8} = \frac{}{8}$	
24.	$2 + \frac{1}{3} = 2\frac{}{3}$	
25.	$\frac{6}{3} + \frac{1}{3} = \frac{}{3}$	
26.	$2 + \frac{1}{3} = \frac{}{3}$	
27.	$2\frac{1}{3} = \frac{}{3}$	
28.	$2 + \frac{1}{5} = 2\frac{}{5}$	
29.	$\frac{10}{5} + \frac{1}{5} = \frac{}{5}$	
30.	$2 + \frac{1}{5} = \frac{}{5}$	
31.	$2\frac{1}{5} = \frac{}{5}$	
32.	$\frac{8}{4} + \frac{3}{4} = \frac{}{4}$	
33.	$2 + \frac{3}{4} = \frac{}{4}$	
34.	$2\frac{3}{4} = \frac{}{4}$	
35.	$\frac{12}{3} + \frac{2}{3} = \frac{}{3}$	
36.	$4 + \frac{2}{3} = \frac{}{3}$	
37.	$4\frac{2}{3} = \frac{}{3}$	
38.	$3 + \frac{3}{5} = \frac{}{5}$	
39.	$3 + \frac{1}{2} = \frac{}{2}$	
40.	$4 + \frac{3}{4} = \frac{}{4}$	
41.	$2 + \frac{1}{6} = \frac{}{6}$	
42.	$2 + \frac{5}{8} = \frac{}{8}$	
43.	$2\frac{4}{5} = \frac{}{5}$	
44.	$3\frac{7}{8} = \frac{}{8}$	

B

Number Correct: ______

Change Fractions to Mixed Numbers

Improvement: ______

1.	$4 + 1 =$	
2.	$\frac{4}{4} + \frac{1}{4} = \frac{}{4}$	
3.	$1 + \frac{1}{4} = \frac{}{4}$	
4.	$1\frac{1}{4} = \frac{}{4}$	
5.	$2 + 1 =$	
6.	$\frac{2}{2} + \frac{1}{2} = \frac{}{2}$	
7.	$1 + \frac{1}{2} = \frac{}{2}$	
8.	$1\frac{1}{2} = \frac{}{2}$	
9.	$5 + 1 =$	
10.	$\frac{5}{5} + \frac{1}{5} = \frac{}{5}$	
11.	$1 + \frac{1}{5} = \frac{}{5}$	
12.	$1\frac{1}{5} = \frac{}{5}$	
13.	$\frac{3}{3} + \frac{1}{3} = \frac{}{3}$	
14.	$1 + \frac{1}{3} = \frac{}{3}$	
15.	$1\frac{1}{3} = \frac{}{3}$	
16.	$1\frac{2}{3} = \frac{}{3}$	
17.	$\frac{10}{10} + \frac{1}{10} = \frac{}{10}$	
18.	$1 + \frac{1}{10} = \frac{}{10}$	
19.	$1\frac{1}{10} = \frac{}{10}$	
20.	$1\frac{7}{10} = \frac{}{10}$	
21.	$\frac{8}{8} + \frac{5}{8} = \frac{}{8}$	
22.	$1 + \frac{5}{8} = \frac{}{8}$	

23.	$1\frac{5}{8} = \frac{}{8}$	
24.	$2 + \frac{1}{2} = 2\frac{}{2}$	
25.	$\frac{4}{2} + \frac{1}{2} = \frac{}{2}$	
26.	$2 + \frac{1}{2} = \frac{}{2}$	
27.	$2\frac{1}{2} = \frac{}{2}$	
28.	$2 + \frac{1}{4} = 2\frac{}{4}$	
29.	$\frac{8}{4} + \frac{1}{4} = \frac{}{4}$	
30.	$2 + \frac{1}{4} = \frac{}{4}$	
31.	$2\frac{1}{4} = \frac{}{4}$	
32.	$\frac{6}{3} + \frac{2}{3} = \frac{}{3}$	
33.	$2 + \frac{2}{3} = \frac{}{3}$	
34.	$2\frac{2}{3} = \frac{}{3}$	
35.	$\frac{12}{4} + \frac{3}{4} = \frac{}{4}$	
36.	$3 + \frac{3}{4} = \frac{}{4}$	
37.	$3\frac{3}{4} = \frac{}{4}$	
38.	$3 + \frac{4}{5} = \frac{}{5}$	
39.	$4 + \frac{1}{2} = \frac{}{2}$	
40.	$4 + \frac{2}{3} = \frac{}{3}$	
41.	$3 + \frac{1}{6} = \frac{}{6}$	
42.	$2 + \frac{7}{8} = \frac{}{8}$	
43.	$2\frac{3}{5} = \frac{}{5}$	
44.	$2\frac{7}{8} = \frac{}{8}$	

A

Number Correct: ______

Change Mixed Numbers to Fractions

1.	$2 + 1 =$	
2.	$\frac{2}{2} + \frac{1}{2} = \frac{}{2}$	
3.	$1 + \frac{1}{2} = \frac{}{2}$	
4.	$1\frac{1}{2} = \frac{}{2}$	
5.	$4 + 1 =$	
6.	$\frac{4}{4} + \frac{1}{4} = \frac{}{4}$	
7.	$1 + \frac{1}{4} = \frac{}{4}$	
8.	$1\frac{1}{4} = \frac{}{4}$	
9.	$3 + 1 =$	
10.	$\frac{3}{3} + \frac{1}{3} = \frac{}{3}$	
11.	$1 + \frac{1}{3} = \frac{}{3}$	
12.	$1\frac{1}{3} = \frac{}{3}$	
13.	$\frac{5}{5} + \frac{1}{5} = \frac{}{5}$	
14.	$1 + \frac{1}{5} = \frac{}{5}$	
15.	$1\frac{1}{5} = \frac{}{5}$	
16.	$1\frac{2}{5} = \frac{}{5}$	
17.	$1\frac{4}{5} = \frac{}{5}$	
18.	$1\frac{3}{5} = \frac{}{5}$	
19.	$\frac{4}{4} + \frac{3}{4} = \frac{}{4}$	
20.	$1 + \frac{3}{4} = \frac{}{4}$	
21.	$\frac{6}{6} + \frac{5}{6} = \frac{}{6}$	
22.	$1 + \frac{5}{6} = \frac{}{6}$	

23.	$1\frac{5}{6} = \frac{}{6}$	
24.	$2 + \frac{1}{2} = 2\frac{}{2}$	
25.	$\frac{4}{2} + \frac{1}{2} = \frac{}{2}$	
26.	$2 + \frac{1}{2} = \frac{}{2}$	
27.	$2\frac{1}{2} = \frac{}{2}$	
28.	$2 + \frac{1}{4} = 2\frac{}{4}$	
29.	$\frac{8}{4} + \frac{1}{4} = \frac{}{4}$	
30.	$2 + \frac{1}{4} = \frac{}{4}$	
31.	$2\frac{1}{4} = \frac{}{4}$	
32.	$\frac{9}{3} + \frac{2}{3} = \frac{}{3}$	
33.	$3 + \frac{2}{3} = \frac{}{3}$	
34.	$3\frac{2}{3} = \frac{}{3}$	
35.	$\frac{16}{4} + \frac{3}{4} = \frac{}{4}$	
36.	$4 + \frac{3}{4} = \frac{}{4}$	
37.	$4\frac{3}{4} = \frac{}{4}$	
38.	$3 + \frac{2}{5} = \frac{}{5}$	
39.	$4 + \frac{1}{2} = \frac{}{2}$	
40.	$3 + \frac{3}{4} = \frac{}{4}$	
41.	$3 + \frac{1}{6} = \frac{}{6}$	
42.	$3 + \frac{5}{8} = \frac{}{8}$	
43.	$3\frac{4}{5} = \frac{}{5}$	
44.	$4\frac{7}{8} = \frac{}{8}$	

B

Number Correct: ______

Change Mixed Numbers to Fractions

Improvement: ______

1.	$5 + 1 =$	
2.	$\frac{5}{5} + \frac{1}{5} = \frac{}{5}$	
3.	$1 + \frac{1}{5} = \frac{}{5}$	
4.	$1\frac{1}{5} = \frac{}{5}$	
5.	$3 + 1 =$	
6.	$\frac{3}{3} + \frac{1}{3} = \frac{}{3}$	
7.	$1 + \frac{1}{3} = \frac{}{3}$	
8.	$1\frac{1}{3} = \frac{}{3}$	
9.	$4 + 1 =$	
10.	$\frac{4}{4} + \frac{1}{4} = \frac{}{4}$	
11.	$1 + \frac{1}{4} = \frac{}{4}$	
12.	$1\frac{1}{4} = \frac{}{4}$	
13.	$\frac{10}{10} + \frac{1}{10} = \frac{}{10}$	
14.	$1 + \frac{1}{10} = \frac{}{10}$	
15.	$1\frac{1}{10} = \frac{}{10}$	
16.	$1\frac{2}{10} = \frac{}{10}$	
17.	$1\frac{4}{10} = \frac{}{10}$	
18.	$1\frac{3}{10} = \frac{}{10}$	
19.	$\frac{3}{3} + \frac{2}{3} = \frac{}{3}$	
20.	$1 + \frac{2}{3} = \frac{}{3}$	
21.	$\frac{8}{8} + \frac{7}{8} = \frac{}{8}$	
22.	$1 + \frac{7}{8} = \frac{}{8}$	

23.	$1\frac{7}{8} = \frac{}{8}$	
24.	$2 + \frac{1}{2} = 2\frac{}{2}$	
25.	$\frac{4}{2} + \frac{1}{2} = \frac{}{2}$	
26.	$2 + \frac{1}{2} = \frac{}{2}$	
27.	$2\frac{1}{2} = \frac{}{2}$	
28.	$2 + \frac{1}{3} = 2\frac{}{3}$	
29.	$\frac{6}{3} + \frac{1}{3} = \frac{}{3}$	
30.	$2 + \frac{1}{3} = \frac{}{3}$	
31.	$2\frac{1}{3} = \frac{}{3}$	
32.	$\frac{12}{4} + \frac{3}{4} = \frac{}{4}$	
33.	$3 + \frac{3}{4} = \frac{}{4}$	
34.	$3\frac{3}{4} = \frac{}{4}$	
35.	$\frac{12}{3} + \frac{2}{3} = \frac{}{3}$	
36.	$4 + \frac{2}{3} = \frac{}{3}$	
37.	$4\frac{2}{3} = \frac{}{3}$	
38.	$3 + \frac{3}{5} = \frac{}{5}$	
39.	$5 + \frac{1}{2} = \frac{}{2}$	
40.	$3 + \frac{2}{3} = \frac{}{3}$	
41.	$3 + \frac{1}{8} = \frac{}{8}$	
42.	$3 + \frac{1}{6} = \frac{}{6}$	
43.	$3\frac{2}{5} = \frac{}{5}$	
44.	$4\frac{5}{6} = \frac{}{6}$	

A

Number Correct: ______

Change Mixed Numbers to Fractions

1.	$4 = 3 + \_\_$	
2.	$\frac{4}{3} = \frac{3}{3} + \frac{}{3}$	
3.	$\frac{4}{3} = 1 + \frac{}{3}$	
4.	$\frac{4}{3} = 1\frac{}{3}$	
5.	$6 = 5 + \_\_$	
6.	$\frac{6}{5} = \frac{5}{5} + \frac{}{5}$	
7.	$\frac{6}{5} = 1 + \frac{}{5}$	
8.	$\frac{6}{5} = 1\frac{}{5}$	
9.	$5 = \_\_ + 1$	
10.	$\frac{5}{4} = \frac{}{4} + \frac{1}{4}$	
11.	$\frac{5}{4} = 1 + \frac{}{4}$	
12.	$\frac{5}{4} = \_\_\ \frac{1}{4}$	
13.	$8 = \_\_ + 3$	
14.	$\frac{8}{5} = \frac{}{5} + \frac{3}{5}$	
15.	$\frac{8}{5} = 1 + \frac{}{5}$	
16.	$\frac{8}{5} = 1\frac{}{5}$	
17.	$\frac{7}{5} = 1\frac{}{5}$	
18.	$\frac{6}{5} = 1\frac{}{5}$	
19.	$\frac{9}{5} = 1\frac{}{5}$	
20.	$\frac{10}{5} =$	
21.	$\frac{}{5} = \frac{10}{5} + \frac{4}{5}$	
22.	$\frac{}{5} = 2 + \frac{4}{5}$	

23.	$\frac{8}{4} =$	
24.	$\frac{}{4} = \frac{8}{4} + \frac{3}{4}$	
25.	$\frac{11}{4} = \frac{8}{4} + \frac{}{4}$	
26.	$\frac{11}{4} = 2 + \frac{}{4}$	
27.	$\frac{11}{4} = 2\frac{}{4}$	
28.	$\frac{}{3} = \frac{6}{3} + \frac{1}{3}$	
29.	$\frac{}{3} = 2 + \frac{1}{3}$	
30.	$\frac{7}{3} = \_\_\ \frac{1}{3}$	
31.	$\frac{8}{3} = \_\_\ \frac{2}{3}$	
32.	$\frac{17}{5} = \frac{}{5} + \frac{2}{5}$	
33.	$\frac{17}{5} = \frac{15}{5} + \frac{}{5}$	
34.	$\frac{17}{5} = \_\_ + \frac{2}{5}$	
35.	$\frac{17}{5} = \_\_\ \frac{2}{5}$	
36.	$\frac{13}{6} = \frac{12}{6} + \frac{}{6}$	
37.	$\frac{13}{6} = \_\_ + \frac{1}{6}$	
38.	$\frac{13}{6} = 2\frac{}{6}$	
39.	$\frac{17}{6} = 2\frac{}{6}$	
40.	$\frac{9}{8} = 1 + \frac{}{8}$	
41.	$\frac{13}{8} = 1 + \frac{}{8}$	
42.	$\frac{19}{10} = 1 + \frac{}{10}$	
43.	$\frac{19}{12} = \frac{}{12} + \frac{7}{12}$	
44.	$\frac{11}{6} = 1 + \frac{}{6}$	

B

Number Correct: ______

Change Mixed Numbers to Fractions

Improvement: ______

1.	$5 = 4 + \_\_$	
2.	$\frac{5}{4} = \frac{4}{4} + \frac{}{4}$	
3.	$\frac{5}{4} = 1 + \frac{}{4}$	
4.	$\frac{5}{4} = 1\frac{}{4}$	
5.	$3 = 2 + \_\_$	
6.	$\frac{3}{2} = \frac{2}{2} + \frac{}{2}$	
7.	$\frac{3}{2} = 1 + \frac{}{2}$	
8.	$\frac{3}{2} = 1\frac{}{2}$	
9.	$9 = \_\_ + 1$	
10.	$\frac{9}{8} = \frac{}{8} + \frac{1}{8}$	
11.	$\frac{9}{8} = 1 + \frac{}{8}$	
12.	$\frac{9}{8} = \_\_\frac{1}{8}$	
13.	$9 = \_\_ + 4$	
14.	$\frac{9}{5} = \frac{}{5} + \frac{4}{5}$	
15.	$\frac{9}{5} = 1 + \frac{}{5}$	
16.	$\frac{9}{5} = 1\frac{}{5}$	
17.	$\frac{8}{5} = 1\frac{}{5}$	
18.	$\frac{7}{5} = 1\frac{}{5}$	
19.	$\frac{6}{5} = 1\frac{}{5}$	
20.	$\frac{8}{4} =$	
21.	$\frac{}{4} = \frac{8}{4} + \frac{1}{4}$	
22.	$\frac{}{4} = 2 + \frac{1}{4}$	
23.	$\frac{6}{3} =$	
24.	$\frac{}{3} = \frac{6}{3} + \frac{2}{3}$	
25.	$\frac{8}{3} = \frac{6}{3} + \frac{}{3}$	
26.	$\frac{8}{3} = 2 + \frac{}{3}$	
27.	$\frac{8}{3} = 2\frac{}{3}$	
28.	$\frac{}{10} = \frac{20}{10} + \frac{1}{10}$	
29.	$\frac{}{10} = 2 + \frac{1}{10}$	
30.	$\frac{21}{10} = \_\_\frac{1}{10}$	
31.	$\frac{27}{10} = \_\_\frac{7}{10}$	
32.	$\frac{13}{6} = \frac{}{6} + \frac{1}{6}$	
33.	$\frac{13}{6} = \frac{12}{6} + \frac{}{6}$	
34.	$\frac{13}{6} = \_\_ + \frac{1}{6}$	
35.	$\frac{13}{6} = \_\_\frac{1}{6}$	
36.	$\frac{17}{8} = \frac{16}{8} + \frac{}{8}$	
37.	$\frac{17}{8} = \frac{}{8} + \frac{1}{8}$	
38.	$\frac{17}{8} = 2\frac{}{8}$	
39.	$\frac{21}{8} = 2\frac{}{8}$	
40.	$\frac{7}{6} = 1 + \frac{}{6}$	
41.	$\frac{11}{6} = 1 + \frac{}{6}$	
42.	$\frac{13}{5} = 2 + \frac{}{5}$	
43.	$\frac{17}{12} = \frac{}{12} + \frac{5}{12}$	
44.	$\frac{13}{8} = 1 + \frac{}{8}$	

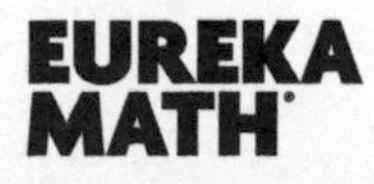

A

Number Correct: ______

Multiply Whole Numbers Times Fractions

1.	$\frac{1}{3} + \frac{1}{3} =$	$\frac{2}{3}$
2.	$2 \times \frac{1}{3} =$	$\frac{2}{3}$
3.	$\frac{1}{4} + \frac{1}{4} + \frac{1}{4} =$	$\frac{3}{4}$
4.	$3 \times \frac{1}{4} =$	$\frac{3}{4}$
5.	$\frac{1}{5} + \frac{1}{5} =$	$\frac{2}{5}$
6.	$2 \times \frac{1}{5} =$	$\frac{2}{5}$
7.	$\frac{1}{5} + \frac{1}{5} + \frac{1}{5} =$	$\frac{3}{5}$
8.	$3 \times \frac{1}{5} =$	$\frac{3}{5}$
9.	$\frac{1}{5} + \frac{1}{5} + \frac{1}{5} + \frac{1}{5} =$	$\frac{4}{5}$
10.	$4 \times \frac{1}{5} =$	$\frac{4}{5}$
11.	$\frac{1}{10} + \frac{1}{10} + \frac{1}{10} =$	$\frac{3}{10}$
12.	$3 \times \frac{1}{10} =$	$\frac{3}{10}$
13.	$\frac{1}{8} + \frac{1}{8} + \frac{1}{8} =$	$\frac{3}{8}$
14.	$3 \times \frac{1}{8} =$	$\frac{3}{8}$
15.	$\frac{1}{2} + \frac{1}{2} =$	$\frac{2}{2}$
16.	$2 \times \frac{1}{2} =$	$\frac{2}{2}$
17.	$\frac{1}{3} + \frac{1}{3} + \frac{1}{3} =$	$\frac{3}{3}$
18.	$3 \times \frac{1}{3} =$	$\frac{3}{3}$
19.	$\frac{1}{4} + \frac{1}{4} + \frac{1}{4} + \frac{1}{4} =$	$\frac{4}{4}$
20.	$4 \times \frac{1}{4} =$	$\frac{4}{4}$
21.	$\frac{1}{2} + \frac{1}{2} + \frac{1}{2} =$	$\frac{3}{2}$
22.	$3 \times \frac{1}{2} =$	$\frac{3}{2}$
23.	$\frac{1}{3} + \frac{1}{3} + \frac{1}{3} + \frac{1}{3} =$	$\frac{4}{3} = 1\frac{1}{3}$
24.	$4 \times \frac{1}{3} =$	$\frac{4}{3} = 1\frac{1}{3}$
25.	$\frac{5}{6} =$	$\frac{4}{6} \times \frac{1}{6}$
26.	$\frac{5}{6} =$	$5 \times$ —
27.	$\frac{5}{8} =$	$5 \times$ —
28.	$\frac{5}{8} =$	— $\times \frac{1}{8}$
29.	$\frac{7}{8} =$	$7 \times$ —
30.	$\frac{7}{10} =$	$7 \times$ —
31.	$\frac{7}{8} =$	— $\times \frac{1}{8}$
32.	$\frac{7}{10} =$	— $\times \frac{1}{10}$
33.	$\frac{6}{6} =$	$6 \times$ __
34.	$1 =$	$6 \times$ __
35.	$\frac{8}{8} =$	— $\times \frac{1}{8}$
36.	$1 =$	— $\times \frac{1}{8}$
37.	$9 \times \frac{1}{10} =$	
38.	$7 \times \frac{1}{5} =$	
39.	$1 =$	$3 \times$ —
40.	$7 \times \frac{1}{12} =$	
41.	$1 =$	— $\times \frac{1}{5}$
42.	$\frac{3}{5} =$	$\frac{1}{5} + \frac{1}{5} +$ —
43.	$3 \times \frac{1}{4} =$	— $+ \frac{1}{4} + \frac{1}{4}$
44.	$1 =$	— + — + —

B

Number Correct: 23

Improvement: ______

Multiply Whole Numbers Times Fractions

1.	$\frac{1}{5}+\frac{1}{5}=$ ✓	$\frac{2}{5}$
2.	$2\times\frac{1}{5}=$ ✓	$\frac{2}{5}$
3.	$\frac{1}{3}+\frac{1}{3}=$ ✓	$\frac{2}{3}$
4.	$2\times\frac{1}{3}=$ ✓	$\frac{2}{3}$
5.	$\frac{1}{4}+\frac{1}{4}+\frac{1}{4}=$ ✓	$\frac{3}{4}$
6.	$3\times\frac{1}{4}=$ ✓	$\frac{3}{4}$
7.	$\frac{1}{5}+\frac{1}{5}+\frac{1}{5}=$ ✓	$\frac{3}{5}$
8.	$3\times\frac{1}{5}=$ ✓	$\frac{3}{5}$
9.	$\frac{1}{5}+\frac{1}{5}+\frac{1}{5}+\frac{1}{5}=$ ✓	$\frac{4}{5}$
10.	$4\times\frac{1}{5}=$ ✓	$\frac{4}{5}$
11.	$\frac{1}{8}+\frac{1}{8}+\frac{1}{8}=$ ✓	$\frac{3}{8}$
12.	$3\times\frac{1}{8}=$ ✓	$\frac{3}{8}$
13.	$\frac{1}{10}+\frac{1}{10}+\frac{1}{10}=$ ✓	$\frac{3}{10}$
14.	$3\times\frac{1}{10}=$ ✓	$\frac{3}{10}$
15.	$\frac{1}{3}+\frac{1}{3}+\frac{1}{3}=$ ✓	$\frac{3}{3}$
16.	$3\times\frac{1}{3}=$ ✓	$\frac{3}{3}$
17.	$\frac{1}{4}+\frac{1}{4}+\frac{1}{4}+\frac{1}{4}=$ ✓	$\frac{4}{4}$
18.	$4\times\frac{1}{4}=$ ✓	$\frac{4}{4}$
19.	$\frac{1}{2}+\frac{1}{2}=$ ✓	$\frac{2}{2}$
20.	$2\times\frac{1}{2}=$ ✓	$\frac{2}{2}$
21.	$\frac{1}{3}+\frac{1}{3}+\frac{1}{3}+\frac{1}{3}=$ ✓	$\frac{4}{3}$
22.	$4\times\frac{1}{3}=$ ✓	$\frac{4}{3}$

23.	$\frac{1}{2}+\frac{1}{2}+\frac{1}{2}=$ ✓	$\frac{3}{2}$
24.	$3\times\frac{1}{2}=$	$\frac{3}{2}$
25.	$\frac{5}{6}=$	$\frac{5}{6}\times\frac{1}{6}$
26.	$\frac{5}{6}=$	$5\times$ —
27.	$\frac{5}{8}=$	$5\times$ —
28.	$\frac{5}{8}=$	— $\times\frac{1}{8}$
29.	$\frac{7}{8}=$	$7\times$ —
30.	$\frac{7}{10}=$	$7\times\frac{10}{10}$
31.	$\frac{7}{8}=$	$\frac{1}{16}\times\frac{1}{8}$
32.	$\frac{7}{10}=$	— $\times\frac{1}{10}$
33.	$\frac{8}{8}=$	$8\times\frac{1}{8}$
34.	$1=$	$8\times\frac{1}{8}$
35.	$\frac{6}{6}=$	$\frac{1}{6}\times\frac{1}{6}$
36.	$1=$	— $\times\frac{1}{6}$
37.	$5\times\frac{1}{12}=$	$\frac{5}{12}$
38.	$6\times\frac{1}{5}=$	$\frac{6}{5}$
39.	$1=$	$4\times\frac{1}{4}$
40.	$9\times\frac{1}{10}=$	$\frac{9}{10}$
41.	$1=$	$\frac{3}{x}\times\frac{1}{3}$
42.	$\frac{3}{4}=$	$\frac{1}{4}+\frac{1}{4}+\frac{1}{4}$
43.	$3\times\frac{1}{5}=$	$\frac{1}{5}+\frac{1}{5}+\frac{1}{5}$
44.	$1=$	$\frac{1}{5}+\frac{1}{5}+\frac{1}{5}+\frac{1}{5}$

Grade 4
Module 6

Todo

A

Number Correct: ______

Divide by 10

1.	20 ÷ 10 =	10
2.	30 ÷ 10 =	10
3.	40 ÷ 10 =	10
4.	80 ÷ 10 =	10
5.	50 ÷ 10 =	10
6.	90 ÷ 10 =	10
7.	70 ÷ 10 =	10
8.	60 ÷ 10 =	10
9.	10 ÷ 10 =	10
10.	100 ÷ 10 =	10
11.	20 ÷ 10 =	10
12.	120 ÷ 10 =	10
13.	50 ÷ 10 =	10
14.	150 ÷ 10 =	10
15.	80 ÷ 10 =	10
16.	180 ÷ 10 =	10
17.	280 ÷ 10 =	10
18.	380 ÷ 10 =	10
19.	680 ÷ 10 =	10
20.	640 ÷ 10 =	64
21.	870 ÷ 10 =	87
22.	430 ÷ 10 =	43
23.	50 ÷ 10 =	5
24.	850 ÷ 10 =	85
25.	1,850 ÷ 10 =	185
26.	70 ÷ 10 =	7
27.	270 ÷ 10 =	27
28.	4,270 ÷ 10 =	427
29.	90 ÷ 10 =	9
30.	590 ÷ 10 =	54
31.	7,590 ÷ 10 =	759
32.	120 ÷ 10 =	12
33.	1,200 ÷ 10 =	120
34.	2,000 ÷ 10 =	
35.	240 ÷ 10 =	
36.	2,400 ÷ 10 =	
37.	4,000 ÷ 10 =	
38.	690 ÷ 10 =	
39.	6,900 ÷ 10 =	
40.	9,000 ÷ 10 =	
41.	940 ÷ 10 =	
42.	5,280 ÷ 10 =	
43.	6,700 ÷ 10 =	
44.	7,000 ÷ 10 =	

B

Number Correct: 15

Divide by 10

Improvement: 20

1.	10 ÷ 10 =	1
2.	20 ÷ 10 =	2
3.	30 ÷ 10 =	3
4.	70 ÷ 10 =	7
5.	40 ÷ 10 =	4
6.	80 ÷ 10 =	8
7.	60 ÷ 10 =	6
8.	50 ÷ 10 =	5
9.	90 ÷ 10 =	9
10.	100 ÷ 10 =	10
11.	30 ÷ 10 =	3
12.	130 ÷ 10 =	30
13.	60 ÷ 10 =	6
14.	160 ÷ 10 =	60
15.	90 ÷ 10 =	9
16.	190 ÷ 10 =	90
17.	290 ÷ 10 =	29
18.	390 ÷ 10 =	39
19.	690 ÷ 10 =	69
20.	650 ÷ 10 =	65
21.	860 ÷ 10 =	86
22.	420 ÷ 10 =	42

23.	40 ÷ 10 =	4
24.	840 ÷ 10 =	84
25.	1,840 ÷ 10 =	184
26.	80 ÷ 10 =	8
27.	280 ÷ 10 =	28
28.	4,280 ÷ 10 =	280
29.	60 ÷ 10 =	6
30.	560 ÷ 10 =	56
31.	7,560 ÷ 10 =	560
32.	130 ÷ 10 =	13
33.	1,300 ÷ 10 =	300
34.	3,000 ÷ 10 =	3,000
35.	250 ÷ 10 =	25
36.	2,500 ÷ 10 =	500
37.	5,000 ÷ 10 =	
38.	740 ÷ 10 =	
39.	7,400 ÷ 10 =	
40.	4,000 ÷ 10 =	
41.	910 ÷ 10 =	
42.	5,820 ÷ 10 =	
43.	7,600 ÷ 10 =	
44.	6,000 ÷ 10 =	

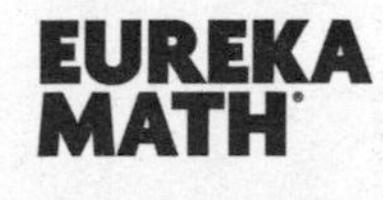

A

Number Correct: ______

Write Fractions and Decimals

1.	$\frac{2}{10}$ =	.	23.	1 =	$\frac{}{10}$
2.	$\frac{3}{10}$ =	.	24.	2 =	$\frac{}{10}$
3.	$\frac{4}{10}$ =	.	25.	5 =	$\frac{}{10}$
4.	$\frac{8}{10}$ =	.	26.	4 =	$\frac{}{10}$
5.	$\frac{6}{10}$ =	.	27.	4.1 =	$\frac{}{10}$
6.	0.1 =	$\frac{}{10}$	28.	4.2 =	$\frac{}{10}$
7.	0.2 =	$\frac{}{10}$	29.	4.6 =	$\frac{}{10}$
8.	0.3 =	$\frac{}{10}$	30.	2.6 =	$\frac{}{10}$
9.	0.7 =	$\frac{}{10}$	31.	3.6 =	$\frac{}{10}$
10.	0.5 =	$\frac{}{10}$	32.	3.4 =	$\frac{}{10}$
11.	$\frac{5}{10}$ =	.	33.	2.3 =	$\frac{}{10}$
12.	0.8 =	$\frac{}{10}$	34.	$4\frac{3}{10}$ =	.
13.	$\frac{7}{10}$ =	.	35.	$\frac{20}{10}$ =	.
14.	0.4 =	$\frac{}{10}$	36.	1.8 =	$\frac{}{10}$
15.	$\frac{9}{10}$ =	.	37.	$3\frac{4}{10}$ =	.
16.	$\frac{10}{10}$ =	.	38.	$\frac{50}{10}$ =	.
17.	$\frac{11}{10}$ =	.	39.	4.7 =	$\frac{}{10}$
18.	$\frac{12}{10}$ =	.	40.	$2\frac{8}{10}$ =	.
19.	$\frac{15}{10}$ =	.	41.	$\frac{30}{10}$ =	.
20.	$\frac{25}{10}$ =	.	42.	3.2 =	$\frac{}{10}$
21.	$\frac{45}{10}$ =	.	43.	$\frac{20}{10}$ =	.
22.	$\frac{38}{10}$ =	.	44.	2.1 =	$\frac{}{10}$

B

Number Correct: ______

Write Fractions and Decimals

Improvement: ______

1.	$\frac{1}{10}$ =	.
2.	$\frac{2}{10}$ =	.
3.	$\frac{3}{10}$ =	.
4.	$\frac{7}{10}$ =	.
5.	$\frac{5}{10}$ =	.
6.	0.2 =	$\frac{\ }{10}$
7.	0.3 =	$\frac{\ }{10}$
8.	0.4 =	$\frac{\ }{10}$
9.	0.8 =	$\frac{\ }{10}$
10.	0.6 =	$\frac{\ }{10}$
11.	$\frac{4}{10}$ =	.
12.	0.9 =	$\frac{\ }{10}$
13.	$\frac{6}{10}$ =	.
14.	0.5 =	$\frac{\ }{10}$
15.	$\frac{9}{10}$ =	.
16.	$\frac{10}{10}$ =	.
17.	$\frac{11}{10}$ =	.
18.	$\frac{12}{10}$ =	.
19.	$\frac{17}{10}$ =	.
20.	$\frac{27}{10}$ =	.
21.	$\frac{47}{10}$ =	.
22.	$\frac{34}{10}$ =	.

23.	1 =	$\frac{\ }{10}$
24.	2 =	$\frac{\ }{10}$
25.	4 =	$\frac{\ }{10}$
26.	3 =	$\frac{\ }{10}$
27.	3.1 =	$\frac{\ }{10}$
28.	3.2 =	$\frac{\ }{10}$
29.	3.6 =	$\frac{\ }{10}$
30.	1.6 =	$\frac{\ }{10}$
31.	2.6 =	$\frac{\ }{10}$
32.	4.2 =	$\frac{\ }{10}$
33.	2.5 =	$\frac{\ }{10}$
34.	$3\frac{4}{10}$ =	.
35.	$\frac{50}{10}$ =	.
36.	1.7 =	$\frac{\ }{10}$
37.	$4\frac{3}{10}$ =	.
38.	$\frac{20}{10}$ =	.
39.	4.6 =	$\frac{\ }{10}$
40.	$2\frac{4}{10}$ =	.
41.	$\frac{40}{10}$ =	.
42.	2.3 =	$\frac{\ }{10}$
43.	$\frac{30}{10}$ =	.
44.	4.1 =	$\frac{\ }{10}$

A

Number Correct: ________

Write Fractions and Decimals

1.	$\frac{3}{10} =$	.	23.	$2 + \frac{1}{10} + \frac{6}{100} =$	.
2.	$\frac{3}{100} =$	.	24.	$2 + 0.1 + 0.06 =$	.
3.	$\frac{23}{100} =$	.	25.	$3 + 0.1 + 0.06 =$	.
4.	$1\frac{23}{100} =$	.	26.	$3 + 0.1 + 0.04 =$	.
5.	$4\frac{23}{100} =$	.	27.	$3 + 0.5 + 0.04 =$	.
6.	$0.07 =$	–	28.	$2 + 0.3 + 0.08 =$	.
7.	$1.07 =$	–	29.	$2 + 0.08 =$	.
8.	$0.7 =$	–	30.	$1 + 0.3 =$	.
9.	$1.7 =$	–	31.	$10 + 0.3 =$	.
10.	$1.74 =$	–	32.	$1 + 0.4 + 0.06 =$	.
11.	$\frac{4}{100} =$	.	33.	$10 + 0.4 + 0.06 =$	.
12.	$0.6 =$	–	34.	$30 + 0.7 + 0.02 =$	.
13.	$\frac{7}{100} =$	.	35.	$2 + \frac{3}{10} + 0.05 =$	.
14.	$0.02 =$	–	36.	$4 + 0.5 + \frac{3}{100} =$	.
15.	$\frac{9}{100} =$	.	37.	$4 + \frac{3}{100} + 0.5 =$	.
16.	$\frac{10}{100} =$	.	38.	$0.5 + \frac{3}{100} + 4 =$	.
17.	$\frac{10}{100} + \frac{2}{100} =$	.	39.	$20 + 0.8 + 0.01 =$	.
18.	$\frac{1}{10} + \frac{2}{100} =$	.	40.	$4 + \frac{9}{100} + \frac{2}{10} =$	.
19.	$\frac{1}{10} + \frac{3}{100} =$	.	41.	$0.04 + 2 + 0.7 =$	–
20.	$\frac{1}{10} + \frac{4}{100} =$	.	42.	$\frac{6}{10} + 8 + \frac{2}{100} =$	.
21.	$\frac{1}{10} + \frac{9}{100} =$	.	43.	$\frac{5}{100} + 8 + 0.9 =$	–
22.	$3 + \frac{1}{10} + \frac{9}{100} =$	.	44.	$0.9 + 10 + \frac{4}{100} =$	.

B

Number Correct: ______

Write Fractions and Decimals

Improvement: ______

1.	$\frac{1}{10} =$	.
2.	$\frac{2}{10} =$	.
3.	$\frac{3}{10} =$	.
4.	$\frac{7}{10} =$	.
5.	$\frac{5}{10} =$	.
6.	0.2 =	–
7.	0.3 =	–
8.	0.4 =	–
9.	0.8 =	–
10.	0.6 =	–
11.	$\frac{4}{10} =$	.
12.	0.9 =	–
13.	$\frac{6}{10} =$	.
14.	0.5 =	–
15.	$\frac{9}{10} =$	.
16.	$\frac{10}{10} =$	.
17.	$\frac{11}{10} =$	.
18.	$\frac{12}{10} =$	.
19.	$\frac{17}{10} =$	.
20.	$\frac{27}{10} =$	.
21.	$\frac{47}{10} =$	.
22.	$\frac{34}{10} =$	.
23.	$2 + \frac{1}{10} + \frac{4}{100} =$	.
24.	2 + 0.1 + 0.04 =	.
25.	3 + 0.1 + 0.04 =	.
26.	3 + 0.1 + 0.06 =	.
27.	3 + 0.5 + 0.06 =	.
28.	2 + 0.4 + 0.09 =	.
29.	2 + 0.06 =	.
30.	1 + 0.5 =	.
31.	10 + 0.5 =	.
32.	1 + 0.2 + 0.04 =	.
33.	10 + 0.2 + 0.04 =	.
34.	30 + 0.9 + 0.06 =	.
35.	$2 + \frac{5}{10} + 0.07 =$	.
36.	$4 + 0.7 + \frac{5}{100} =$	.
37.	$4 + \frac{5}{100} + 0.7 =$	.
38.	$0.7 + \frac{5}{100} + 4 =$	.
39.	20 + 0.6 + 0.01 =	.
40.	$6 + \frac{7}{100} + \frac{4}{10} =$	.
41.	0.06 + 2 + 0.9 =	–
42.	$\frac{8}{10} + 6 + \frac{4}{100} =$	.
43.	$\frac{3}{100} + 8 + 0.7 =$	–
44.	$0.7 + 10 + \frac{6}{100} =$	.

A

Number Correct: ________

Add Decimal Fractions

1.	$\frac{1}{10} =$	.
2.	$\frac{1}{100} =$	.
3.	$\frac{1}{10} + \frac{1}{100} =$	.
4.	$\frac{3}{10} =$	.
5.	$\frac{3}{100} =$	.
6.	$\frac{3}{10} + \frac{3}{100} =$	.
7.	$\frac{5}{10} =$	.
8.	$\frac{5}{100} =$	.
9.	$\frac{5}{10} + \frac{5}{100} =$	.
10.	$\frac{7}{10} =$	.
11.	$\frac{9}{100} =$	.
12.	$\frac{7}{10} + \frac{9}{100} =$	.
13.	$\frac{9}{100} + \frac{7}{10} =$	.
14.	$\frac{4}{10} =$	.
15.	$\frac{6}{100} =$	.
16.	$\frac{4}{10} + \frac{6}{100} =$	.
17.	$\frac{4}{100} + \frac{6}{10} =$	.
18.	$\frac{8}{10} + \frac{5}{100} =$	.
19.	$\frac{9}{10} + \frac{2}{100} =$	.
20.	$\frac{1}{100} + \frac{8}{10} =$	.
21.	$\frac{4}{100} + \frac{1}{10} =$	.
22.	$\frac{7}{100} + \frac{4}{10} =$	.

23.	$\frac{2}{10} =$	.
24.	$\frac{20}{100} =$	.
25.	$\frac{2}{10} + \frac{20}{100} =$	.
26.	$\frac{3}{10} =$	.
27.	$\frac{30}{100} =$	.
28.	$\frac{3}{10} + \frac{30}{100} =$	.
29.	$\frac{5}{10} + \frac{20}{100} =$	.
30.	$\frac{8}{10} + \frac{10}{100} =$	.
31.	$\frac{8}{10} + \frac{20}{100} =$	.
32.	$\frac{8}{10} + \frac{30}{100} =$	.
33.	$\frac{8}{10} + \frac{50}{100} =$	.
34.	$\frac{9}{10} + \frac{40}{100} =$	.
35.	$\frac{9}{10} + \frac{47}{100} =$	.
36.	$\frac{7}{10} + \frac{50}{100} =$	.
37.	$\frac{7}{10} + \frac{59}{100} =$	.
38.	$\frac{6}{10} + \frac{60}{100} =$	.
39.	$\frac{6}{10} + \frac{64}{100} =$	.
40.	$\frac{65}{100} + \frac{6}{10} =$	.
41.	$\frac{91}{100} + \frac{7}{10} =$	.
42.	$\frac{8}{10} + \frac{73}{100} =$	.
43.	$\frac{9}{10} + \frac{82}{100} =$	.
44.	$\frac{98}{100} + \frac{9}{10} =$	.

B

Number Correct: _______

Add Decimal Fractions

Improvement: _______

1.	$\frac{2}{10} =$	.
2.	$\frac{2}{100} =$	.
3.	$\frac{2}{10} + \frac{2}{100} =$	.
4.	$\frac{4}{10} =$	.
5.	$\frac{4}{100} =$	.
6.	$\frac{4}{10} + \frac{4}{100} =$	.
7.	$\frac{6}{10} =$	.
8.	$\frac{6}{100} =$	.
9.	$\frac{6}{10} + \frac{6}{100} =$	.
10.	$\frac{4}{10} =$	.
11.	$\frac{8}{100} =$	.
12.	$\frac{4}{10} + \frac{8}{100} =$	.
13.	$\frac{8}{100} + \frac{4}{10} =$	.
14.	$\frac{5}{10} =$	.
15.	$\frac{7}{100} =$	.
16.	$\frac{5}{10} + \frac{7}{100} =$	.
17.	$\frac{7}{100} + \frac{5}{10} =$	.
18.	$\frac{9}{10} + \frac{6}{100} =$	.
19.	$\frac{8}{10} + \frac{3}{100} =$	.
20.	$\frac{1}{100} + \frac{7}{10} =$	.
21.	$\frac{3}{100} + \frac{1}{10} =$	.
22.	$\frac{8}{100} + \frac{3}{10} =$	.
23.	$\frac{1}{10} =$	.
24.	$\frac{10}{100} =$	.
25.	$\frac{1}{10} + \frac{10}{100} =$	.
26.	$\frac{4}{10} =$	.
27.	$\frac{40}{100} =$	.
28.	$\frac{4}{10} + \frac{40}{100} =$	.
29.	$\frac{5}{10} + \frac{30}{100} =$	.
30.	$\frac{7}{10} + \frac{20}{100} =$	.
31.	$\frac{7}{10} + \frac{30}{100} =$	.
32.	$\frac{7}{10} + \frac{40}{100} =$	.
33.	$\frac{7}{10} + \frac{60}{100} =$	.
34.	$\frac{9}{10} + \frac{30}{100} =$	.
35.	$\frac{9}{10} + \frac{37}{100} =$	.
36.	$\frac{8}{10} + \frac{40}{100} =$	.
37.	$\frac{8}{10} + \frac{49}{100} =$	.
38.	$\frac{7}{10} + \frac{70}{100} =$	.
39.	$\frac{7}{10} + \frac{76}{100} =$	.
40.	$\frac{78}{100} + \frac{7}{10} =$	.
41.	$\frac{81}{100} + \frac{7}{10} =$	.
42.	$\frac{9}{10} + \frac{73}{100} =$	.
43.	$\frac{9}{10} + \frac{84}{100} =$	.
44.	$\frac{84}{100} + \frac{8}{10} =$	.

Grade 4
Module 7

A

Number Correct: ______

Convert to Dollars

1.	1 cent =	$ 0.
2.	2 cents =	
3.	3 cents =	
4.	8 cents =	
5.	80 cents =	
6.	70 cents =	
7.	60 cents =	
8.	20 cents =	
9.	1 penny =	
10.	1 dime =	
11.	2 pennies =	
12.	2 dimes =	
13.	3 pennies =	
14.	3 dimes =	
15.	9 dimes =	
16.	7 pennies =	
17.	8 dimes =	
18.	4 pennies =	
19.	6 dimes =	
20.	8 pennies =	
21.	7 dimes =	
22.	9 pennies =	

23.	6 pennies =	
24.	5 dimes =	
25.	5 pennies =	
26.	1 dime 1 penny =	
27.	1 dime 2 pennies =	
28.	1 dime 7 pennies =	
29.	4 dimes 5 pennies =	
30.	6 dimes 3 pennies =	
31.	3 pennies 6 dimes =	
32.	7 pennies 9 dimes =	
33.	1 quarter =	
34.	2 quarters =	
35.	3 quarters =	
36.	2 quarters 3 pennies =	
37.	1 quarter 3 pennies =	
38.	3 quarters 3 pennies =	
39.	2 quarters 2 dimes =	
40.	1 quarter 1 dime =	
41.	3 quarters 1 dime =	
42.	1 quarter 4 dimes =	
43.	3 quarters 2 dimes =	
44.	3 quarters 18 pennies =	

B

Number Correct: _______

Improvement: _______

Convert to Dollars

1.	2 cents =	$ 0.	23.	5 pennies =	
2.	3 cents =		24.	6 dimes =	
3.	4 cents =		25.	4 pennies =	
4.	9 cents =		26.	1 dime 1 penny =	
5.	90 cents =		27.	1 dime 2 pennies =	
6.	80 cents =		28.	1 dime 8 pennies =	
7.	70 cents =		29.	5 dimes 4 pennies =	
8.	30 cents =		30.	7 dimes 4 pennies =	
9.	1 penny =		31.	4 pennies 7 dimes =	
10.	1 dime =		32.	6 pennies 8 dimes =	
11.	2 pennies =		33.	1 quarter =	
12.	2 dimes =		34.	2 quarters =	
13.	3 pennies =		35.	3 quarters =	
14.	3 dimes =		36.	2 quarters 4 pennies =	
15.	8 dimes =		37.	1 quarter 4 pennies =	
16.	6 pennies =		38.	3 quarters 4 pennies =	
17.	7 dimes =		39.	2 quarters 3 dimes =	
18.	9 pennies =		40.	1 quarter 2 dimes =	
19.	5 dimes =		41.	3 quarters 2 dimes =	
20.	7 pennies =		42.	1 quarter 5 dimes =	
21.	9 dimes =		43.	3 quarters 1 dime =	
22.	8 pennies =		44.	3 quarters 19 pennies =	

Name ____________________ Date ____________

Practice Set A Part 1: Multi-Digit Addition Fluency

1.
$$8,149 + 7,264$$

2.
$$42,609 + 8,685$$

3.
$$39,563 + 48,438$$

4.
$$658,199 + 25,675$$

5.
$$445,976 + 37,415$$

6.
$$438,617 + 493,859$$

Practice Set A Part 2: Multi-Digit Addition Fluency

1.
$$9,202 + 6,211$$

2.
$$42,774 + 8,520$$

3.
$$53,545 + 34,456$$

4.
$$604,754 + 79,120$$

5.
$$454,315 + 29,076$$

6.
$$110,728 + 821,748$$

Name ______________________________ Date ____________________

Practice Set B Part 1: Multi-Digit Subtraction Fluency

1.

$$\begin{array}{r} 7,739 \\ -\ 5,546 \\ \hline \end{array}$$

2.

$$\begin{array}{r} 23,145 \\ -\ 5,129 \\ \hline \end{array}$$

3.

$$\begin{array}{r} 71,378 \\ -\ 61,876 \\ \hline \end{array}$$

4.

$$\begin{array}{r} 479,541 \\ -\ 78,856 \\ \hline \end{array}$$

Practice Set B Part 2: Multi-Digit Subtraction Fluency

1.

$$\begin{array}{r} 7,699 \\ -\ 5,506 \\ \hline \end{array}$$

2.

$$\begin{array}{r} 19,145 \\ -\ 1,129 \\ \hline \end{array}$$

3.

$$\begin{array}{r} 71,878 \\ -\ 62,376 \\ \hline \end{array}$$

4.

$$\begin{array}{r} 479,497 \\ -\ 78,812 \\ \hline \end{array}$$

Name ______________________________ Date ________________

Practice Set C Part 1: Multi-Digit Subtraction with Zeros Fluency

1.

$$\begin{array}{r} 7{,}890 \\ -\ 5{,}472 \\ \hline \end{array}$$

2.

$$\begin{array}{r} 28{,}001 \\ -\ 5{,}853 \\ \hline \end{array}$$

3.

$$\begin{array}{r} 60{,}407 \\ -\ 35{,}344 \\ \hline \end{array}$$

4.

$$\begin{array}{r} 400{,}069 \\ -\ 24{,}362 \\ \hline \end{array}$$

Practice Set C Part 2: Multi-Digit Subtraction with Zeros Fluency

1.

$$\begin{array}{r} 7{,}890 \\ -\ 5{,}472 \\ \hline \end{array}$$

2.

$$\begin{array}{r} 28{,}609 \\ -\ 6{,}461 \\ \hline \end{array}$$

3.

$$\begin{array}{r} 60{,}497 \\ -\ 35{,}434 \\ \hline \end{array}$$

4.

$$\begin{array}{r} 400{,}869 \\ -\ 25{,}162 \\ \hline \end{array}$$

Name ______________________________ Date ________________

Practice Set D Part 1: Multi-Digit Addition and Subtraction Fluency

1.
$$\begin{array}{r} 9{,}327 \\ +\ 9{,}664 \\ \hline \end{array}$$

2.
$$\begin{array}{r} 39{,}463 \\ -\ 38{,}938 \\ \hline \end{array}$$

3.
$$\begin{array}{r} 758{,}194 \\ +\ 35{,}478 \\ \hline \end{array}$$

4.
$$\begin{array}{r} 839{,}014 \\ -\ 27{,}075 \\ \hline \end{array}$$

5.
$$\begin{array}{r} 438{,}615 \\ +\ 193{,}979 \\ \hline \end{array}$$

6.
$$\begin{array}{r} 960{,}043 \\ -\ 368{,}972 \\ \hline \end{array}$$

Practice Set D Part 2: Multi-Digit Addition and Subtraction Fluency

1.
$$\begin{array}{r} 9{,}630 \\ +\ 9{,}361 \\ \hline \end{array}$$

2.
$$\begin{array}{r} 34{,}478 \\ -\ 33{,}953 \\ \hline \end{array}$$

3.
$$\begin{array}{r} 754{,}454 \\ +\ 39{,}218 \\ \hline \end{array}$$

4.
$$\begin{array}{r} 839{,}099 \\ -\ 27{,}160 \\ \hline \end{array}$$

5.
$$\begin{array}{r} 108{,}215 \\ +\ 524{,}379 \\ \hline \end{array}$$

6.
$$\begin{array}{r} 959{,}943 \\ -\ 368{,}872 \\ \hline \end{array}$$

A

Number Correct: ______

Convert Length Units

1.	1 km =	m
2.	2 km =	m
3.	3 km =	m
4.	7 km =	m
5.	5 km =	m
6.	1 m =	cm
7.	2 m =	cm
8.	3 m =	cm
9.	9 m =	cm
10.	6 m =	cm
11.	1 yd =	ft
12.	2 yd =	ft
13.	3 yd =	ft
14.	10 yd =	ft
15.	5 yd =	ft
16.	1 ft =	in
17.	2 ft =	in
18.	3 ft =	in
19.	10 ft =	in
20.	4 ft =	in
21.	9 km =	m
22.	4 km =	m

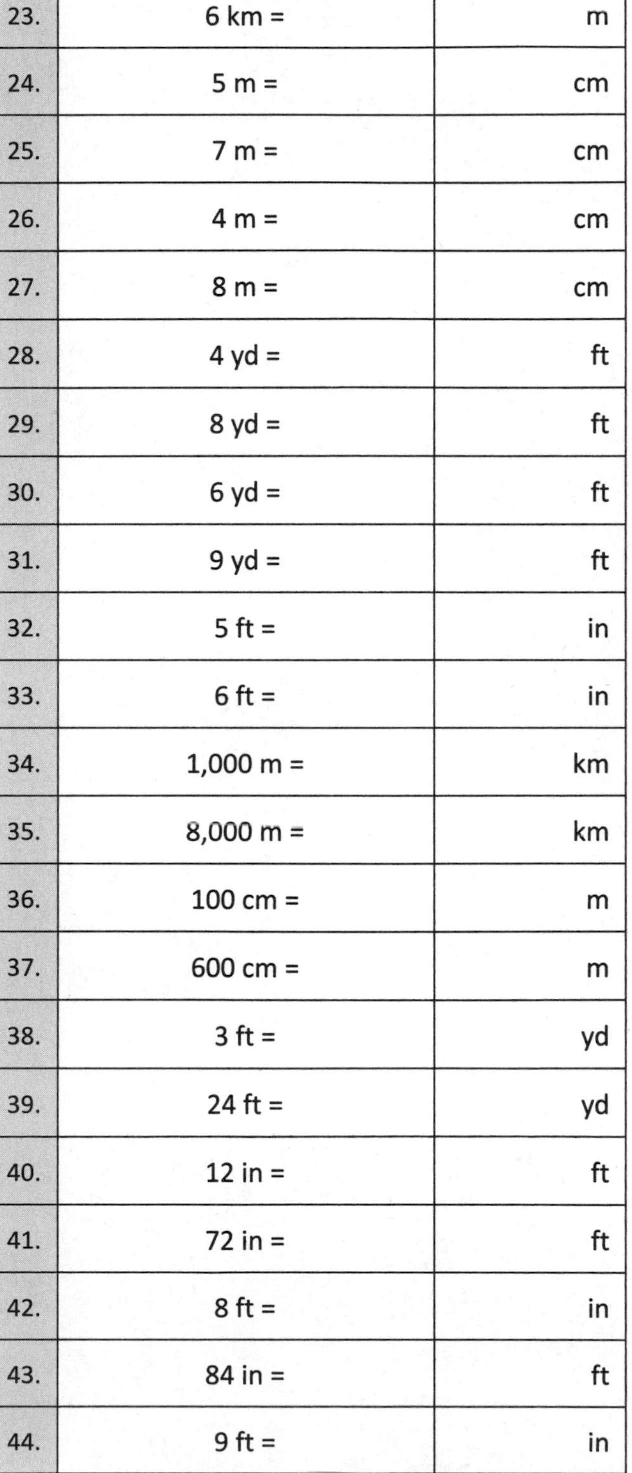

23.	6 km =	m
24.	5 m =	cm
25.	7 m =	cm
26.	4 m =	cm
27.	8 m =	cm
28.	4 yd =	ft
29.	8 yd =	ft
30.	6 yd =	ft
31.	9 yd =	ft
32.	5 ft =	in
33.	6 ft =	in
34.	1,000 m =	km
35.	8,000 m =	km
36.	100 cm =	m
37.	600 cm =	m
38.	3 ft =	yd
39.	24 ft =	yd
40.	12 in =	ft
41.	72 in =	ft
42.	8 ft =	in
43.	84 in =	ft
44.	9 ft =	in

B

Number Correct: ______

Convert Length Units

Improvement: ______

1.	1 m =	cm	23.	6 m =	cm
2.	2 m =	cm	24.	5 km =	m
3.	3 m =	cm	25.	7 km =	m
4.	7 m =	cm	26.	4 km =	m
5.	5 m =	cm	27.	8 km =	m
6.	1 km =	m	28.	6 yd =	ft
7.	2 km =	m	29.	9 yd =	ft
8.	3 km =	m	30.	4 yd =	ft
9.	9 km =	m	31.	8 yd =	ft
10.	6 km =	m	32.	5 ft =	in
11.	1 yd =	ft	33.	6 ft =	in
12.	2 yd =	ft	34.	100 cm =	m
13.	3 yd =	ft	35.	800 cm =	m
14.	5 yd =	ft	36.	1,000 m =	km
15.	10 yd =	ft	37.	6,000 m =	km
16.	1 ft =	in	38.	3 ft =	yd
17.	2 ft =	in	39.	27 ft =	yd
18.	3 ft =	in	40.	12 in =	ft
19.	10 ft =	in	41.	84 in =	ft
20.	4 ft =	in	42.	9 ft =	in
21.	9 m =	cm	43.	72 in =	ft
22.	4 m =	cm	44.	8 ft =	in

Credits

Great Minds® has made every effort to obtain permission for the reprinting of all copyrighted material. If any owner of copyrighted material is not acknowledged herein, please contact Great Minds for proper acknowledgment in all future editions and reprints of this module.